SpringerBriefs in Computer Science

Series Editors

Stan Zdonik
Peng Ning
Shashi Shekhar
Jonathan Katz
Xindong Wu
Lakhmi C. Jain
David Padua
Xuemin Shen
Borko Furht
V. S. Subrahmanian
Martial Hebert
Katsushi Ikeuchi
Bruno Siciliano

For further volumes:
http://www.springer.com/series/10028

K. K. Shukla · Arvind K. Tiwari

Efficient Algorithms for Discrete Wavelet Transform

With Applications to Denoising and Fuzzy Inference Systems

 Springer

K. K. Shukla
Banaras Hindu University
Indian Institute of Technology
Varanasi
Uttar Pradesh
India

Arvind K. Tiwari
GE India Technology Center
Bangalore
India

ISSN 2191-5768   ISSN 2191-5776 (electronic)
ISBN 978-1-4471-4940-8   ISBN 978-1-4471-4941-5 (eBook)
DOI 10.1007/978-1-4471-4941-5
Springer London Heidelberg New York Dordrecht

Library of Congress Control Number: 2013930141

Printed on acid-free paper

Springer is part of Springer Science+Business Media (www.springer.com)

Preface

Transforms occupy an important compartment of an engineer's tool kit for solving signal processing and polynomial computation problems efficiently. By resolving a time-varying signal into sinusoidal components, engineers transform a problem from that of studying time domain phenomena to that of evaluating frequency domain properties. These properties often lead to simple explanations of otherwise complicated occurrences. Further, polynomial arithmetic can be implemented efficiently in the transform domain.

In contrast to the Fourier transform based approaches where a fixed window is used uniformly for a spread of frequencies, the wavelet transform uses short windows at high frequencies and long windows at low frequencies. In this way, the characteristics of non-stationary disturbances can be more closely monitored. In other words, both time and frequency information can be obtained by wavelet transform. Instead of transforming a pure *time description* into a pure *frequency description*, the wavelet transform finds a good promise in a *time-frequency* description.

Due to its inherent timescale locality characteristics, the discrete wavelet transform (DWT) has received considerable attention in digital signal processing (speech and image processing), communication, computer science, and mathematics. Wavelet transforms are known to have excellent energy compaction characteristics and are able to provide perfect reconstruction. Therefore, they are ideal for signal/image processing. The shifting (or translation) and scaling (or dilation) are unique to wavelets. Orthogonality of wavelets with respect to dilations leads to multi-grid representation.

The nature of wavelet computation forces us to carefully examine the implementation methodologies. As the computation of DWT involves filtering, an efficient filtering process is essential in DWT hardware implementation. In the multistage DWT, coefficients are calculated recursively, and in addition to the wavelet decomposition stage, extra space is required to store the intermediate coefficients. Hence, the overall performance depends significantly on the precision of the intermediate DWT coefficients.

The book presents new sequential and parallel implementation techniques of DWT that are efficient. Efficiency of proposed techniques is in terms of computation requirement, storage requirement, and reconstructed signal with better signal-to-noise ratio. Applications to signal denoising power quality prediction are discussed.

Contents

Symbols and Notations

σ^2	Noise variance
$\eta_{mth}(.)$	Modified thresholding function
$\hat{f}$	Reconstructed test function
$\eta_{th}(.)$	Thresholding function
φ	Scaling function
ψ	Wavelet function
λ	Gain parameter of modified thresholding function
ε	Noise
τ	Translation parameter
ω_c	Filter cut off frequency
$*$	Complex conjugate
B	Word length in fixed point representation
db4	Daubechies filter (Tap length 4)
F	Original test function
G	Analysis low pass filter
G'	Synthesis low pass filter
H	Analysis high pass filter
H'	Synthesis high pass filter
L	Filter tap length
N_f	Number of fractional bits
N_i	Number of integer bits
S	Scale (dilation) parameter
Sym8	Symmlet filter (Tap length 8)
Z	Set of integers

Chapter 1
Introduction

Abstract *Wavelet transforms* (WT) have growing impact on *signal processing* theory and practice. This is because of two reasons: (a) unifying role of wavelet transform and (b) their successes in wide variety of applications. *Digital filter banks*, the basis of *wavelet*-based algorithms, have become standard signal processing operators. Filter banks are the fundamental tools required for processing of real signals using digital signal processors (DSP) [133,139]. Vaidyanathan in his book [134] has discussed connection between theory of filter bank and DSP. The purpose of this book is to look at wavelet-related issues from a signal processing perspective. This book focuses on and around new implementation techniques of discrete wavelet transform (DWT) and their applications in denoising and classification. On this account, it is required to introduce the wavelet theory in brief. The organization of this chapter is as follows: Section 1.1 introduces the subject in brief. Section 1.2 presents historical review of multiresolution analysis and wavelet transform. Various kinds of wavelet transform applied to signal processing applications viz. continuous wavelet transform (CWT) and DWT (one dimension and two dimensions) are discussed in brief. Section 1.3 reviews implementation issues and applications of DWT from signal processing viewpoint. Section 1.4 concludes this chapter by outlining major contribution of the book.

Keywords Wavelet transform · Denoising · Multiresolution analysis · Digital filter · SNR

1.1 Wavelet Transform

Multiresolution analysis of signals and phenomena has been a topic of interest for researchers from wide variety of fields. Prominent areas of application are summarized by Daubechies et al. [34]. Some of them are given in Table 1.1.

K. K. Shukla and A. K. Tiwari, *Efficient Algorithms for Discrete Wavelet Transform*,   1
SpringerBriefs in Computer Science, DOI: 10.1007/978-1-4471-4941-5_1,
© K. K. Shukla 2013

Table 1.1 Prominent areas of application of wavelet transform

Discipline	Applications
Maths	Harmonic analysis
Physics	Phase space analysis
Digital signal processing	Multirate filtering, quadrature mirror filters, and sub-band coding
Image processing	Pyramidal image representation and compression
Speech processing	Efficient representation and equalization

Thus, it is clear that the wavelet analysis is useful for problems in many disciplines. Therefore, it is obvious that there is something special about it. Resinkoff et al. in their book [115] had rightly proposed wavelets as new mathematical engineering. Wavelet analysis provides a systematic new way to represent and analyze multiscale structures. Natural structures and engineering problems are closer to multiscale nature, which is one reason why wavelets are broadly valuable. Wavelet analysis is also a far-reaching generalization of orthogonal bases of functions. Wavelets are a systematic way to represent functions on unbounded domains by linear combinations of orthogonal basis functions that are *compactly supported* and *overlapped*. These basis functions are potentially realizable by physical devices [115].

From an historical viewpoint, wavelet theory has been developed only recently, although similar ideas and constructions date back to the researchers of the nineteenth century [107, 137]. Fourier laid the foundations with his theories of frequency analysis, which proved to be enormously important and influential. The attention of researchers gradually turned from frequency-based analysis to scale-based analysis when it started to become clear that an approach measuring average fluctuations at different scales might prove less sensitive to noise. Morlet and the team at the *Marseille Theoretical Physics Center* working under Alex Grossmann in France first proposed the concept of wavelets in its present theoretical form [59]. Meyer et al. [96], who have ensured the methods' dissemination, have developed the methods of wavelet analysis. The main algorithm dates back to the work of Mallat [85, 86]. Since then, research on wavelets has become international. Contributions of the work of scientists such as Daubechies [33–37], Coifman, and Wickerhauser [29] have made wavelet theory popular in researchers from various disciplines of science and engineering. The wavelet domain is growing up very quickly. A lot of mathematical papers and practical trials published every month are available online www.wavelet.org.

Wavelet analysis presents an efficient representation for a wide class of functions. The class of functions that can be represented by wavelet analysis is much larger than the class of square-integrable functions for finite energy signals. The wavelet analysis and efficient representation of functions bear an implicit relationship. This forces researchers to think for solution of a class of operational and philosophical questions as follows:

How to search for efficient solutions of large and complex practical engineering problems?

The *state of the art* of computing systems largely influences the solution of a complex problem. Further, in part, it also depends to a large degree on the efficiency of its mathematical representation and on the rate of convergence of the mathematical processes that are employed to solve the problem [115]. Following are the areas of importance to be singled out where wavelets have a major role to play:

- Efficient algorithm for computation of wavelet transform so that results are more accurate with optimum hardware resources.
- Signal denoising algorithm is based on the wavelet expansion representation in a manner that systematically reduces noise, so that the resulting signal (image, audio signal etc.) is a better representation of the information than the given signal was in its original form.
- Faster computation of wavelet coefficients.

1.2 Multiresolution Representations and Wavelets

In the seminal paper by Daubechies [36], compactly supported wavelets were first introduced. Multiresolution analysis introduced by Mallat [87] used Daubechies wavelets as a critical building block and explored the massive representational power possessed by these new square-integrable functions. Oliver et al. [107] introduced wavelet theory as unified framework for various signal processing applications [36, 53, 87, 96, 107, 132].

Multiresolution representation is a new term for a very old idea. It organizes information into categories called *levels* and usually arranges it so that *higher* in hierarchy a level is, the fewer the coefficients it has. A multiresolution structure provides different ways of grouping things to reveal aspects of structure that depends on the *scale* of activity. In biological realm, the human vision system employs several multiresolution structures. One *design objective* of the vision system is to provide wide-aperture detection (so events can be detected early) and high-resolution detection (so that the detailed structure of the visual event can be seen) [115]. Since the vision system has limited bandwidth, the *uncertainty principle* tells us that these objectives are fundamentally incompatible. Nature has evolved a multiresolution solution, which allocates the limited available bandwidth in two parts:

(a) The bulk of the retinal receptors is arranged as a wide-aperture but low-activity sensor (the "principal part"),
(b) Whereas, a small fraction of the sensors form the fovea, which has a much higher resolution but a narrow aperture (the "residual part").

This system provides multiresolution information because, after detecting peripheral motion, *we turn our gaze* to see the details. The vision system trades time for bandwidth. This is a classical engineering trade-off that must be made in many problems.

The concept of *multiresolution analysis* underlies the theory of wavelets. The idea is simple and ancient: The information to be analyzed is separated into a *principal* part and a *residual* part. In applications to signal processing, the principal part should be thought of as primarily *low pass* and the residual part as primarily *high pass*. The reason for this identification is that there are more *high-frequency* states than *low-frequency* states, so reduction in complexity amounts to selection of a suitable *low-pass* principal. Figure 1.1 plots fundamental cell of multiresolution analysis.

Further, question arises what makes a function a wavelet and why wavelets are desirable in certain applications. The term *wavelet* as it implies means a little wave. This little wave has at least one oscillation in both the positive and negative directions and a fast decay to zero. This property is analogous to an admissibility condition of a function that is required for wavelet transform [113]. Sets of wavelets are employed to approximate a signal. The goal is to find a set of daughter wavelets constructed by *dilation* (scaled or compressed) and *translation* (shifted) of original wavelet (mother wavelet), which represents the signal efficiently. Thus, by *traveling* from the large scales toward the finer scales, one *zooms in* and arrives at more and more exact representations of given signal.

Given that the wavelet field keeps growing, the definition of wavelet continuously changes. Therefore, it is almost impossible to define a wavelet perfectly. Sweldens in his work [144] has enumerated rightly what is currently meant by wavelet. The definition of wavelet with notations and symbols as presented in [144] is as follows:

A wavelet is denoted as $\psi_\lambda(x)$ where x belongs to the (undefined) spatial domain X, ψ_λ belongs to a (undefined) class F of functions, and λ belongs to an (undefined) index set Λ.

Given X is the real line, F as $L^2(R)$, Λ as $\mathbf{Z}^2$ with $\lambda = (j, l)$,

and $\psi_\lambda(x) = 2^{-j/2}\psi_\lambda(2^j x - l)$,

where $\psi = \{\psi_\lambda | \lambda \in \Lambda\}$ is referred to as the wavelet basis.

The types of wavelet transform and their application vary with nature of problem under investigation. For a continuous input signal, the time-scale parameters can be continuous leading to a CWT [87]. Its discrete version leads to *Wavelet Series* expansion [36, 87, 96]. The wavelet transform for discrete time signals is defined as DWT [36, 87, 96]. It is worth to note its analogy with (continuous) Fourier transform, Fourier series, and discrete Fourier transform (DFT).

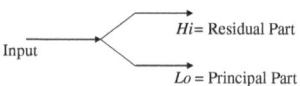

Fig. 1.1 Fundamental cell of a multiresolution analysis (*MRA*)

1.2.1 Continuous-Time Wavelets

Consider a real or complex value continuous-time function $\psi(t)$, which denotes a vector space for finite energy signals and satisfies

$$\int_{-\infty}^{+\infty} |\psi(t)|^2 dt \langle \infty. \tag{1.1a}$$

A restriction on the choice of $\psi(t)$ is that it must have a zero average value and be of short duration, which mathematically is called the admissibility condition on $\psi(t)$ [37] given by

$$\int_{-\infty}^{\infty} \psi(t) dt = 0 \tag{1.1b}$$

The CWT of $x(t)$ (square-integrable function) [113] is defined as

$$\text{CWT}(s,t) = \int_{-\infty}^{+\infty} x(t).\psi_{s,\tau}^*(t) dt \langle \infty \tag{1.2}$$

where

$$\psi_{s,\tau}(t) = |s|^{-\frac{1}{2}} \psi\left(\frac{t-\tau}{s}\right) \tag{1.3}$$

$\psi(t)$ is the basis function or the *mother wavelet* which defines the details, * denotes a complex conjugate, and $s, \tau \in R$ are the dilation and translation parameters, respectively ($s \neq 0$).

The CWT maps an one-dimensional function to function of two continuous real variables s and τ. The region of support of CWT(s,τ) is defined as the set of ordered pairs (s,τ) for which CWT$(s,\tau) \neq 0$. The CWT provides a redundant representation of signal in the sense that the entire support of CWT$(s,)$ need not be used to recover the function under investigation.

Wavelets are families of functions generated from one single function, called an *analyzing wavelet* or *mother wavelet*, by means of scaling and translating operations. Plots of some useful mother wavelets are shown in Fig. 1.2. The difference between these wavelets is mainly due to the different lengths of filters that define the wavelet and scaling functions. Wavelets must be oscillatory, must decay quickly to zero (can only be nonzero for a short period), and must integrate to zero.

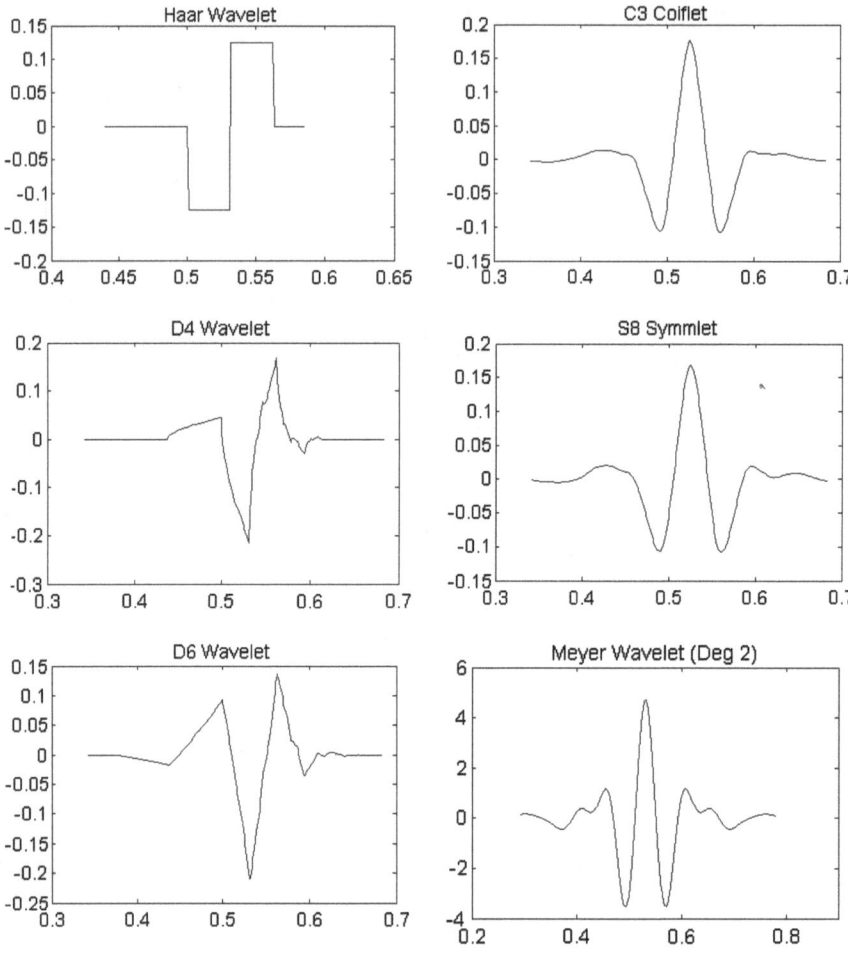

Fig. 1.2 Mother wavelets often used in wavelet analysis (MATLAB generated [14])

1.2.2 Discretization of Time-Scale Parameter

The CWT introduces wavelet functions whose behavior is similar to an orthonormal basis. In practical implementation, all the real-world signals are band limited. Thus, discrete dilation and translation parameters, maintaining orthonormal basis, will be of use instead of its continuous version. Oliver et al. [107], while describing wavelet frames and orthonormal bases, presented a discretized time-scale parameter of CWT. There is a natural way to discretize it [37]. The two scales $s_0 < s_1$ roughly correspond to two frequencies $f_0 > f_1$. The wavelet coefficients at scale s_1 can be subsampled at (f_0/f_1)th rate of the coefficients at scale s_0 according to *Nyquist* rules. Thus, the time-scale parameter is discretized on the sampling grid as shown in Fig. 1.3. Here, the relation between subsequent levels of

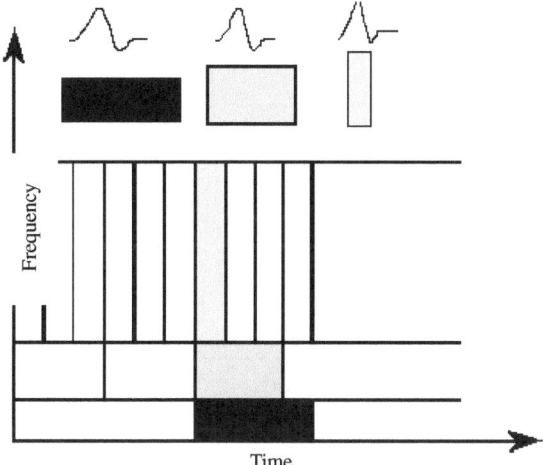

Fig. 1.3 Dyadic sampling grid in the time-scale plane

scale and time parameters will be given by $s = s_0^j$ and $\tau = k\tau_0 s_0^j$, where $j, k \in Z$. Here, s_0 and τ_0 are fixed constant with $s_0 \geq 1$ and $\tau_0 \geq 0$, and the corresponding discrete mother wavelets are [37, 87, 107]

$$\psi_{j,k}(t) = s_0^{-j/2} \psi\left(\frac{t - k\tau s_0^j}{s_0^j}\right) \tag{1.4}$$

and resultant discretized wavelet coefficients are given by

$$DWT_{j,k} = \int x(t)\psi_{j,k}^*(t)dt \tag{1.5}$$

By careful selection of s_0 and τ_0, the family of dilated *mother wavelets* can constitute an orthonormal basis of $L^2(R)$. The simplest choice of s_0 and τ_0 is $s_0 = 2$ and $\tau_0 = 1$, resulting in a *dyadic-orthonormal wavelet transform* [87]. Figure 1.4a plots translated and dilated versions of Daubechies (db3) wavelet. Figure 1.4b is a plot of Symlet (sym8) [14] wavelet at different scales and translations.

1.2.3 Discrete Wavelet Transform

Computation of wavelet coefficients at every possible scale is a fair amount of work, and it generates an awful lot of data. Selection of a subset of scales and positions based on powers of two (dyadic scales and positions) results in a more efficient and accurate analysis. Mallat [87] has introduced repetitive application of *low-pass* and *high-pass* filters to calculate the wavelet expansion of a given sequence of discrete numbers as plotted in Fig. 1.5 [37, 87, 107, 137].

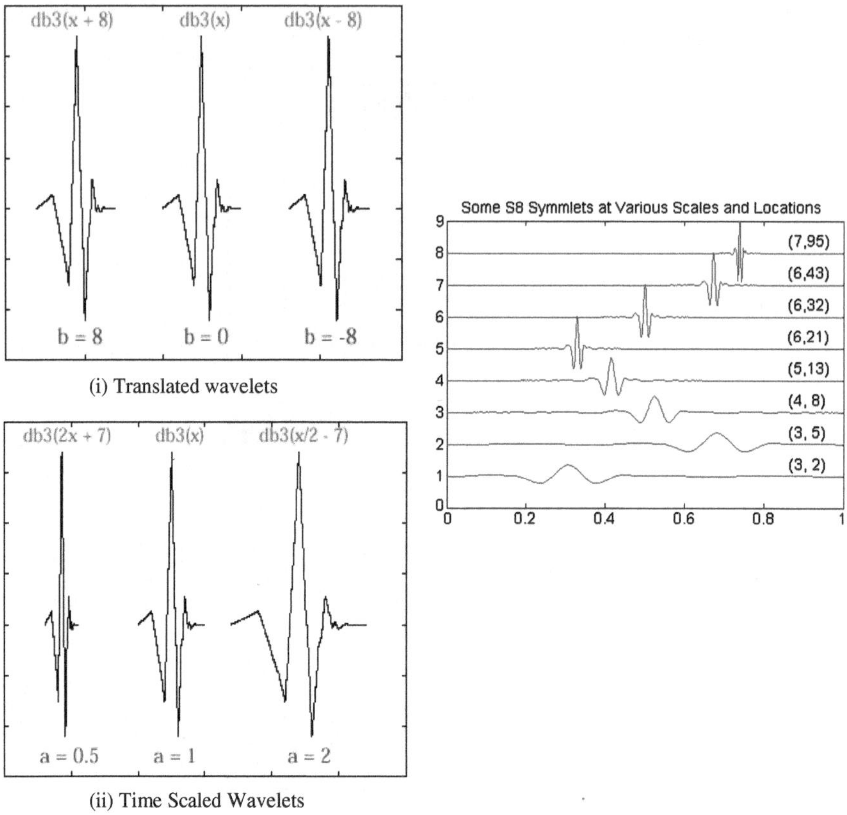

db3(x + 8)	db3(x)	db3(x - 8)

Some S8 Symmlets at Various Scales and Locations

b = 8   b = 0   b = -8

(i) Translated wavelets

db3(2x + 7)	db3(x)	db3(x/2 - 7)

a = 0.5   a = 1   a = 2

(ii) Time Scaled Wavelets

Fig. 1.4 **a** Translated and dilated versions of Daubechies (db3) wavelet (MATLAB generated [14]). **b** Symlet (sym8) wavelet at different scales and translations (MATLAB generated [14])

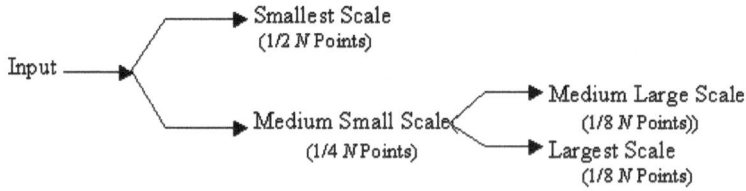

Fig. 1.5 Mallat's cascaded filter MRA: tree structure DWT

Vetterli [137] has presented the approximation properties of filter banks and their relation to wavelets in his paper. An orthonormal compactly supported wavelet basis of $L^2(R)$ is formed by the dilation and translation of a single real-valued function $\psi(t)$, called the *mother wavelet* given by

$$\psi_{j,k}(t) = 2^{-j/2}\psi\left(\frac{t - k2^j}{2^j}\right); \quad j, k \in Z, \tag{1.6}$$

where Z is the set of integers. In Eq. (1.6), the function ψ has M vanishing moments [37] up to order $(M - 1)$ and it satisfies the following *two-scale difference equation*,

$$\psi_{j,k}(t) = \sqrt{2} \sum_{k=0}^{L-1} h_k.\varphi(2t - k) \tag{1.7}$$

where $\varphi(t)$ is a companion function of the *wavelet* function, which is called *scaling function*, used to define the approximations and forms a set of orthonormal bases of $L^2(R)$ as given below:

$$\varphi_{j,k}(t) = 2^{-j/2}\varphi\left(\frac{t - k2^j}{2^j}\right); \quad j, k \in Z \tag{1.8}$$

The *scaling function* $\varphi(t)$ satisfies

$$\int_{-\infty}^{+\infty} \varphi(t).dt = 1 \tag{1.9}$$

and two-*scale difference equation*

$$\varphi_{j,k}(t) = \sqrt{2} \sum_{k=0}^{L-1} g_k.\varphi(2t - k) \tag{1.10}$$

The wavelet transform computation requires a pair of filters. One filter in the pair calculates the wavelet coefficients, whereas the other applies the scaling function. This scaling function, implemented with filter coefficients $\{g_k\}$, provides an approximation of the signal via the following equations [87]:

$$W_L(n,j) = \sum_m W_L(m, j - 1)g(m - 2n) \tag{1.11a}$$

The wavelet function gives us the detail signals which are also called high-pass output as [87]:

$$W_H(n,j) = \sum_m W_L(m, j - 1)h(m - 2n) \tag{1.11b}$$

where $W_L(p,q)$ is the pth scaling coefficient at the qth stage, $W_H(p,q)$ is the pth wavelet coefficient at the qth stage, and $g(n)$, $h(n)$ are the filter coefficients corresponding to the scaling (*low-pass* filter) and wavelet (*high-pass* filter) functions, respectively [137]. These two filters are related by

$$h_k = (-1)^k.g_{L-k}; \quad k = 0, \ldots, L - 1 \tag{1.12}$$

and are called *quadrature mirror filters* (QMF) [137].

The above equations demonstrate that the computation of DWT can be performed using FIR filters. The transform is recursive. The *low-pass* outputs are used to compute the wavelet coefficients at the next octave (level of resolution). The DWT is a nonredundant type wavelet representation. The two-dimensional sequence $d_{j,k}$ is commonly referred to as the DWT [113]. The *DWT* is still the transform of a continuous-time signal. The discretization is only in scale and translation parameters $(s,)$. The algorithm gets its efficiency by halving the output data at every resolution known as subsampling. The algorithm has a complexity of $O(N)$. For more details on wavelet theory, readers are advised to see [107].

1.2.4 Wavelet Reconstruction

Mallat [87] in his paper has shown to obtain the original discrete sequence by a pyramid transform. The process of starting with a sequence of the approximation coefficients at some level of resolution and then generating the approximation and detail coefficients at coarser levels through digital filtering and decimation is referred to as a *decomposition* or *analysis* of the sequence. The reverse process of combining the coarser approximation and detail coefficients to yield the approximation coefficients at a finer resolution, performed by digital filtering, is referred to as *reconstruction* or *synthesis*. The mathematical manipulation that affects synthesis is called the *inverse discrete wavelet transform* (IDWT) [87, 108, 109, 113].

The filtering part of the reconstruction process also bears some discussion, because it is the choice of filters that is crucial in achieving perfect reconstruction of the original signal. The downsampling of the signal components, performed during the decomposition phase, introduces a distortion called aliasing. It turns out that by carefully choosing filters for the decomposition and reconstruction phases, which are closely related (but not identical), the effect of aliasing can be canceled.

Using basis functions, $\varphi(t)$ and $\psi(t)$, and it can be derived that the original function $x(t)$ is represented by

$$x(t) = \sum_k C_{n_0,k} \cdot \varphi_{n_0,k}(t) \; + \; \sum_{j=1}^{n_0} \sum_k d_{j,k} \cdot \psi_{j,k}(t) \tag{1.13}$$

where $\sum_k C_{n_0,k} \cdot \varphi_{n_0,k}(t)$ is the smooth or approximate representation of the original signal at a resolution $j = n_0$ ($n_0 > 0$ and an integer) and $\sum_{j=1}^{n_0} \sum_k d_{j,k} \cdot \psi_{j,k}(t)$ are the n_0 detailed representations at resolution $1, 2, \ldots, n_0$. The approximations are *high-scale, low-frequency* components of the signal. The details are the *low-scale, high-frequency* components. In terms of filter bank to reconstruct the original data, the DWT coefficients are *upsampled* [87, 108, 109] and passed through another set of low- and high-pass filters, which is expressed as follows:

$$W_L(n,j) = \sum_k W_L(k,j+1)g'(n-2k) + \sum_l W_H(l,j+1)h'(n-2l) \tag{1.14}$$

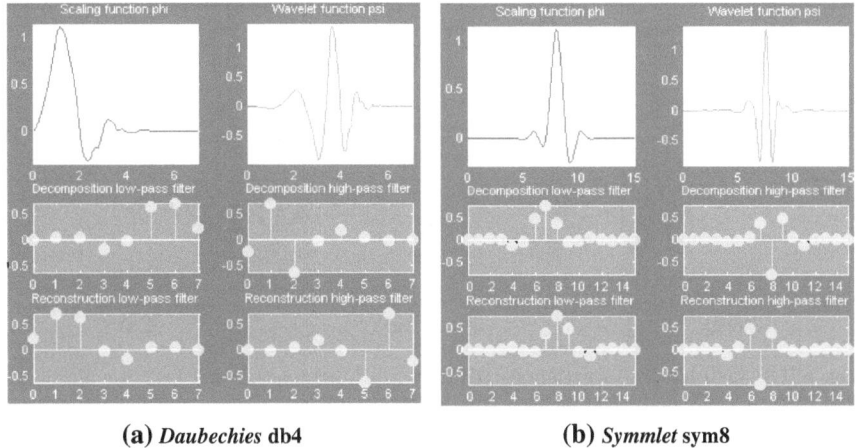

(a) *Daubechies* db4   **(b)** *Symmlet* sym8

Fig. 1.6 Impulse response plots of decomposition and reconstruction filters corresponding to scaling and wavelet functions (MATLAB generated [14]). **a** *Daubechies* db4. **b** *Symlet* sym8

where $g'(n)$ and $h'(n)$ are, respectively, the *low-pass* and *high-pass* synthesis filter corresponding to the mother wavelet. Figure 1.6 plots impulse response of decomposition and reconstruction filters corresponding to scaling and wavelet functions of *Daubechies* (db4) and *Symlet* (sym8) wavelets [14].

1.2.5 Multistep Decomposition and Reconstruction

Figure 1.7 represents a multistep analysis/synthesis DWT process. This involves two aspects: breaking up a signal to obtain the wavelet coefficients and reassembling the signal from the coefficients. We have already discussed decomposition and reconstruction at some length. Of course, there is no point in breaking up a signal merely to have the satisfaction of immediately reconstructing it. The wavelet coefficients may get modified before performing the reconstruction step [87, 113, 107].

1.2.6 Two-Dimensional DWT

DWT decomposition of multidimensional signals (e.g., images) necessitates multidimensional extensions of wavelets. It is easier to generalize the one-dimensional DWT model to any dimension $n > 0$ given by Mallat [87]. Products of wavelets and scaling functions corresponding to 1D DWT produce two-dimensional separable wavelets [138]. This section presents two-dimensional

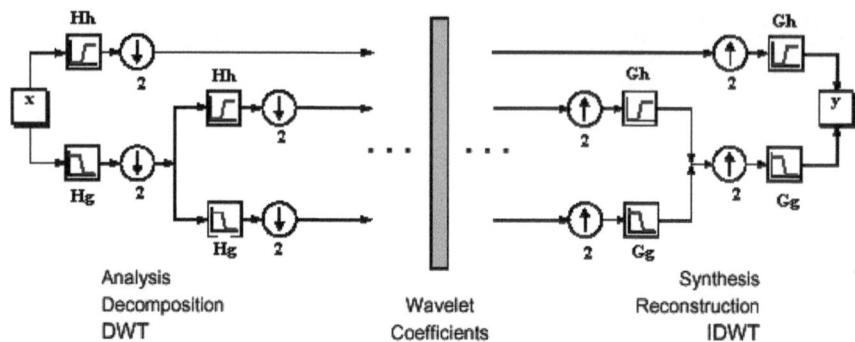

Fig. 1.7 A multistep analysis/synthesis DWT (*H* and *G* are *analysis* and *synthesis* filter bank, respectively; *h* and *g* are *high*- and *low*-pass filters, respectively)

DWT decomposition, which will be of use in image processing applications. The signal $f(x, y) \in L^2(R^2)$ is a square-integrable, finite energy function of two variables x and y. A sequence of subspaces of $L^2(R^2)$ is a *multiresolution* approximation of $L^2(R^2)$. Let $\psi(x)$ (1.6) and $\varphi(x)$ (1.8) are the one-dimensional wavelet and scaling functions, respectively. Corresponding two-dimensional scaling function can be generated as follows [37, 87, 107, 113, 138]:

$$\varphi(x, y) = \varphi(x)\varphi(y) \tag{1.15}$$

and associated three wavelets are

$$\psi_1(x, y) = \varphi(x)\psi(y) \tag{1.16}$$

$$\psi_2(x, y) = \psi(x)\varphi(y) \tag{1.17}$$

$$\psi_3(x, y) = \psi(x)\psi(y) \tag{1.18}$$

The generated scaling and wavelet functions are orthogonal to each other with respect to integer shifts. The scaling function $\varphi(x, y)$ is a *low-pass* filter, whereas $\psi_i(x, y)_{,i=1,2,3}$ wavelet functions are corresponding *high-pass* filters. These correspond to two-dimensional filter bank structure with subsampling by 2 in each dimension, resulting in overall subsampling by 4, Fig. 1.8.

The total number of pixels in transform domain representation is equal to the number of pixels of the original image, so the process does not increase volume of data in transform domain. In two dimensions, the wavelet representation is computed with pyramidal algorithm similar to one dimension. It computes wavelet transform along the x and y-axis. Further, one-dimensional reconstruction algorithm is applicable to two-dimensional case also. Figure 1.9 plots the four functions associated with the two-dimensional Symlet (sym8) [14] wavelet.

Fig. 1.8 Two-dimensional
DWT decomposition [87].
a Filter bank representation.
b The partition of frequency
plane

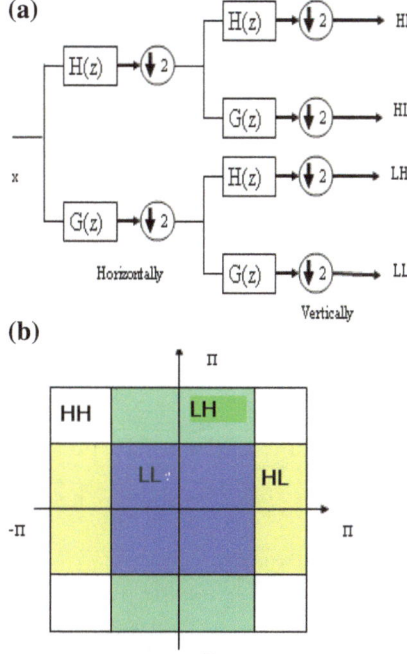

1.3 Review of DWT Implementation Issues and Applications

Real-life data instead of being totally random have certain correlation structure, for example, audio signals, images, solutions of differential equations, time series. The correlation structure of many of these signals is similar. They have some correlation with space (or time), but the correlation is local. For example, neighboring pixels in an image are highly correlated, but ones that are far from each other are uncorrelated. Similarly, there is some correlation with frequency, but again it is local. Indeed, the spectrum of many signals has a band structure [137, 144].

This motivates approximating real-life data sets with building blocks with space and frequency localization properties. The power of such building blocks lies in their ability to reveal the internal correlation structure of the data sets. This result in powerful approximation qualities: *only a small number of building blocks may provide an accurate approximation of the data.*

Selection of basis functions to solve a problem dates back to Fourier and his investigation of heat equation [137]. In short, a basis function is chosen both for its ability to represent an object of interest (for example, good approximation with few coefficients) and for its operational value [137, 153].

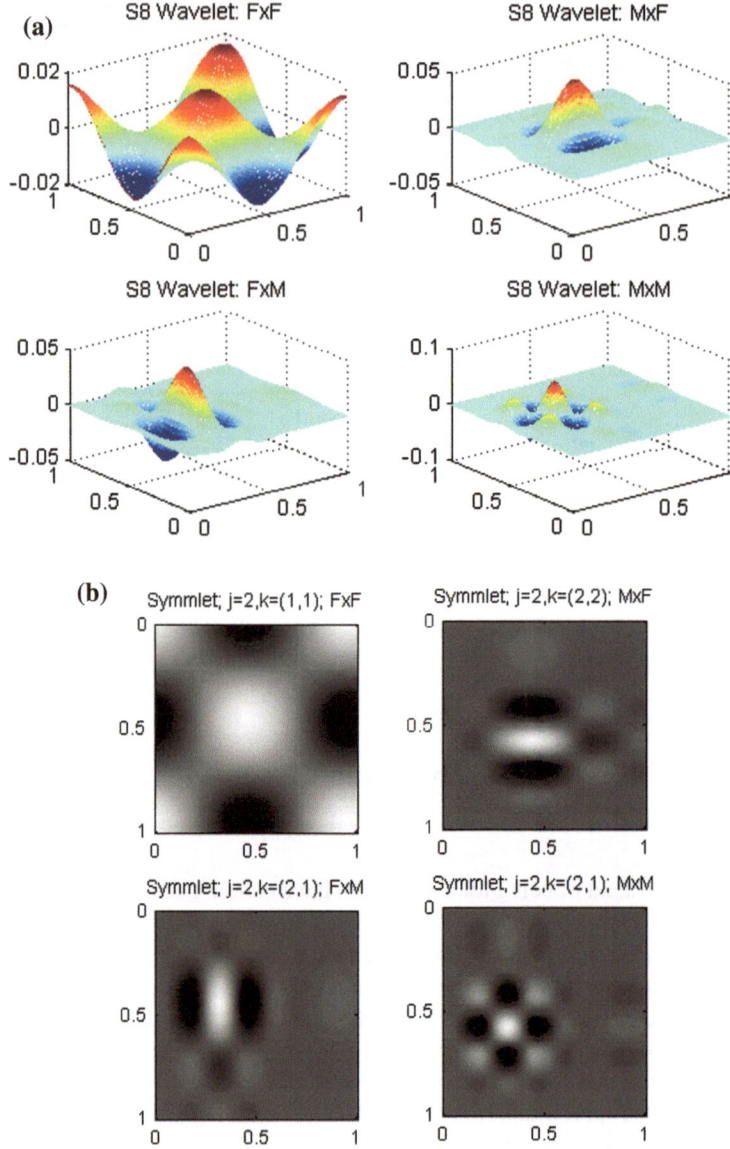

Fig. 1.9 a Two-dimensional symlet (sym8) wavelet—mesh plots (MATLAB generated [14], F: φ, M: ψ). **b** Two-dimensional Symlet (sym8) wavelet—image plots (MATLAB generated [14], F: φ, M: ψ)

Let S be, for example, space of integrable functions f with finite square integral

$$\int |f(t)|^2 \mathrm{d}t \langle \infty \tag{1.19}$$

Let any element $f \in S$ can be written as a linear combination

$$f = \sum_{i \in I} \alpha_i \varphi_i \qquad (1.20)$$

Further, consider $\hat{f}_N(t)$ is the approximate representation of f using N terms, to be chosen appropriately given by

$$\hat{f}_N = \sum_N \alpha_n \hat{\varphi}_n \qquad (1.21)$$

The L_2 norm of the approximation error is

$$\left\| f(t) - \hat{f}_N(t) \right\|_2 = \left(\int\limits_{-\infty}^{+\infty} \left| f(t) - \hat{f}_N(t) \right|^2 dt \right)^{\frac{1}{2}} \qquad (1.22)$$

The approximation is said to be good provided

$$\lim_{N \to \infty} \left\| f(t) - \hat{f}_N(t) \right\|_2 = 0 \qquad (1.23)$$

This necessitates investigation of various issues related to representation of $f(t)$ satisfying (1.23). There are various ways to choose N terms used in the approximation. A fixed subset (e.g., the first N terms) leads to a linear, subspace approximation. Adaptive methods, to be discussed later, are nonlinear. For example, denoising in wavelet bases has led to interesting results for piecewise smooth signals precisely because of the superior approximation properties for such signals. The function approximation involves not only approximation quality, but also representation cost. There is a cost associated with describing $\hat{f}_N(t)$, and this cost depends on the approximation method and operation involved in processing. Typically, here the fixed-point representation of function in wavelet domain is investigated. Further, to reduce the operational cost in processing, a parallel digital filter representation of wavelet structure is explored. To improve the performance, parallel implementation of wavelet algorithms in parallel virtual machine (PVM) [55] environment is presented.

1.3.1 Digital Filter Bank and Finite Word-Length Effects

As discussed, an efficient way to implement the DWT is using digital filters. The Mallat's algorithm [87] is in fact a classical scheme known in the signal processing community as a *two-channel sub-band coder* [127]. Gadre et al. [53] has discussed effect of finite word length on stability of multirate filters. Authors in their paper have reported that it is necessary to use about 18- to 20-bit accuracy for an order 16 *low-pass* filter and 10- to 12-bit accuracy for order 8 filters [53, 87, 108, 127].

In practice, the data under processing are represented with a finite word length. The result of processing leads to numbers requiring additional bits for their representation. For example, a *b-bit* data sample multiplied by a *b-bit* coefficient results in a product that is *2b bits* long. It can be followed that if the results of arithmetic operations are not quantized in recursive realization of digital filter, the number of bits will increase indefinitely. The effect of quantization in such applications depends on factors such as whether *fixed-point* numbers represent *fraction* or *integers* and whether *rounding or truncation* performs *quantization* [108]. For *fixed-point* arithmetic, it is natural in a signal processing context to consider a register as representing a fixed-point fraction. In this way, the product of two numbers remains a fraction and truncating or rounding the least-significant bits can maintain the limited register length. In this type of representation, the result of addition of *fixed-point* fractions need not be truncated or rounded. However, the magnitude of resulting sum can exceed unity. This effect is commonly referred to as *overflow* and can be handled by proper *scaling* of input data. Effect of truncation or rounding of results in an arithmetic operation is manifested in the form of inserting nonlinearity in the filter. Thus, precise analysis of truncation or rounding errors in DWT analysis/synthesis is required from practical implementation viewpoint. A common objective of error analysis is to choose the register length necessary to meet some specifications on the relative sizes of signal and errors. The register length can, of course, be changed only in steps of one bit. As will be discussed in this chapter, the addition of one bit to the register length reduces the amplitude of quantization errors by a factor of approximately one-half. Thus, a final decision concerning register length is insensitive to inaccuracies in the error analysis. An analysis correct to within 30–40 % is often adequate [108]. Because of this insensitivity, in present investigation, a statistical model of DWT is proposed to estimate the quantization errors in final results due to fixed-point implementation.

An error analysis reveals that when performed on fixed-point arithmetic units, the usual Mallat's pyramid algorithm [87] looses number of bits. This phenomenon, often referred to as computational noise, is caused by inaccurate computations due to inaccurate limited dynamic range. Long DWTs are unavoidable in high-resolution analysis of signals (such as shock waves in oil search, or acoustic echoes in imaging radars). Therefore, it is common to switch to custom hardware based on floating-point arithmetic. Unfortunately, floating-point integrated circuits are more complex and expensive than fixed-point processors. Any improvement of fixed-point computation so as to be more robust and accurate could lead to significantly less-expensive chips in high-precision DWT processors.

1.3.2 Discrete Wavelet Transform and Signal Denoising

There have been vast investigation into removal of noise in signals and images using wavelet transform. The principal work is that of Donoho and Johnstone

[39–42] based on thresholding of wavelet coefficients and reconstructing it. The method relies on the fact that in wavelet domain, noise tends to be represented by wavelet coefficients at finer scales [113]. As the coefficients at such scales are also the primary carriers of edge information, it requires developing a methodology for the removal of these coefficients. The removal must be in order that one looses minimum of primary signal components and at the same time removes the noise to a maximum [39–42, 87, 113, 137].

1.3.2.1 Function Approximation and Wavelets

Selection of *wavelet function* primarily depends on its ability to represent an object of interest (for example, good approximation with few coefficients). This immediately necessitates answer to a number of different questions. Different *wavelet function* can give very different rates of approximation [137]. There are various ways to select the number of terms used in approximation.

The signal denoising application is a special case of function estimation (from finite samples). Wavelet thresholding is an approach of selecting orthogonal basis functions. Under this setting, linear coefficients for each basis function can be readily estimated (due to orthogonality), and one can construct a sparse code by an appropriate ordering of the coefficients (i.e., by their magnitude), followed by thresholding or discarding insignificant or small coefficients. Considering the following model of a discrete noisy signal:

$$y_i = f(t_i) + \varepsilon_i, \quad i = 1, 2, \dots, N \tag{1.24}$$

In what follows, suppose the samples are equally spaced in the unit interval [0,1], then the sample size N is power of two: $N = 2^J$, for some positive integer J. These assumptions allow us to perform both the DWT and IDWT using Mallat's algorithm [87]. Due to the orthogonality of the matrix W, the DWT of white noise is also an array of independent $N(0,1)$ random variables, so from (1.24), it follows that

$$\hat{c}_{j_0 k} = c_{j_0 k} + \varepsilon_{jk} \tag{1.25}$$

$$\hat{d}_{jk} = d_{j_0 k} + \varepsilon_{jk} \tag{1.26}$$

where, $k = 0, 1, \dots 2^{j_0} - 1$, $j = j_0, \dots (J - 1)$ and $\hat{c}_{j_0 k}$ and $\hat{d}_{j_0 k}$ are, respectively, the empirical scaling and the empirical wavelet coefficients of the signal under consideration. The sparseness of the wavelet expansion makes it reasonable to assume that essentially only a few *large* d_{jk} contain information about the underlying function f, while *small* d_{jk} can be attributed to the noise, which uniformly contaminates all wavelet coefficients. Whether it can be decided which are the significant large wavelet coefficients that will be retained and, setting all others equal to zero, result in an approximate wavelet representation of the signal is under investigation. It is also advisable to keep the scaling coefficients, $c_{j_0 k}$ the

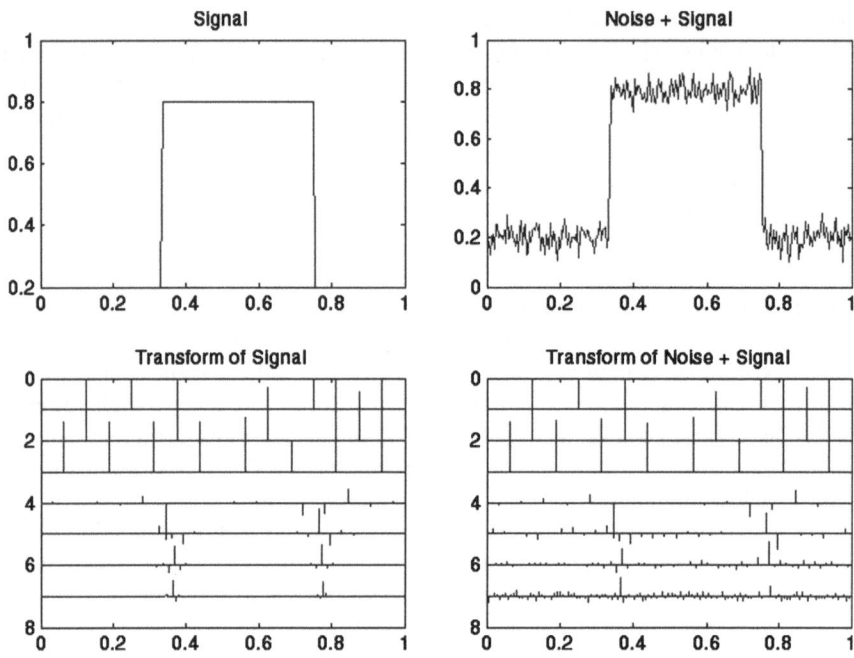

Fig. 1.10 Motivation for denoising: a *square* pulse function and its corrupted version along with their wavelet decompositions. (MATLAB generated)

coefficients on the lower *coarse* levels, intact because they represent *low-frequency* terms that usually contain important information about the signal.

To strengthen the above discussion, consider the example in Fig. 1.10, where a square pulse function has been corrupted by additive noise and the goal is to recover the original function. The wavelet coefficients of the original function and the noisy function are displayed in Fig 1.10. It is observed that at coarser level, wavelet decomposition of both original function and its counterpart has less number of coefficients of comparatively higher magnitude. Also, it is clear from decomposition of noisy function that at higher level of decompositions, it has more number of coefficients in comparison with original function of small magnitude. Thus, it is clear that additive noise results in wavelet decomposition in the form of coefficients of small magnitude at higher level (i.e., corresponding to high-frequency components).

Recovery of original function in wavelet domain is possible by setting a threshold value, which sets the coefficients corresponding to noise to zero in wavelet domain. Hence, the question becomes

1. How to distinguish between the coefficients that are mainly due to signal and those mainly due to noise.
2. How should the thresholds be adjusted?

1.3.3 Fast Computation of Discrete Wavelet Transform

Digital signal processing applications involve use of transforms invariably. Practical implementation of these transforms is computationally intensive, and complicated data manipulation is required. Recently, users from both high-performance scientific community and general-purpose applications have shown keen interest in parallel processing due its higher performance, lower cost, and sustained productivity [118]. To solve a computationally intensive problem efficiently on a cluster of existing computers (distributed computing) involves a significantly lower cost factor. Although it is difficult for a distributed—computing user to achieve the computational capacity of large massively parallel processors (MPP)—it is possible to solve large-size problems by combining a variety of distributed computing resources, connected by high-speed networks. This approach has advantages in terms of flexibility, stability, scalability, and low costs. The advantage of using a cluster of workstations as the computational platform is that a cluster of a large number of workstations is easily available. A disadvantage is that there may be many users running unrelated tasks on the workstations so that the available computing resource for each task fluctuates in an unpredictable manner. Furthermore, communication between workstations is relatively slow [49–51, 60, 107, 118, 156].

The implementation of DWT is computationally intensive [107, 60, 49–51]. The parallel implementation of DWT algorithm has been proposed on a cluster of distributed computing system in PVM environment. In contrast to the traditional parallel approaches, which rely on specialized parallel machines, this work explores the potential of distributed systems for parallelism. The *master/slave* model is adopted for control of machines. It has been observed that parallelism provides more powerful computing ability for DWT decomposition and reconstruction on test *signals/images* with large data sets [156]. The suitability of this approach lies in the parallel implementation of DWT on a low-cost heterogeneous PVM network.

1.3.4 Discrete Wavelet Transform Applied to Power Quality Signal Classification

The power quality study involves an important step, that is, monitoring the actual voltage and current waveforms and classifies and displays these when certain thresholds are exceeded. The extensive usage of power electronics devices for conditioning and control of electric power by both the consumers and electric utilities has resulted in more of power quality problems. The classification scheme has to be robust and accurate to handle the noisy data collected from the transmission and distribution networks. Santoso et al. in their work [119–121] have shown capabilities of wavelet transform in detection and classification of power quality problems. Kulkarni et al. [78] have developed a new transform for detection of single as well as multiple transient signals [5, 34, 78, 119–121, 154, 155].

This book explores application of DWT as a powerful tool for detection, localization, classification, and quantification of power signal disturbances. The techniques discussed deal with the problem in wavelet domain, which covers both time and frequency domains simultaneously. It generates a time–frequency picture of the signal distorted due to different classes of faults [154]. The distribution of the energy of the distorted signal at different resolution levels has been used as input to fuzzy logic block for classification of different disturbances. The DWT has been used as preprocessor, and afterward, processing of wavelet coefficients by fuzzy logic has been carried out [155]. Wavelet preprocessing reduces the dimensionality of the problem. In this book, a comparative study of various defuzzification schemes in connection to fuzzy inferencing system has been carried out [157].

With this understanding, the book presents new implementation techniques of DWT that are efficient. Efficiency of proposed technique is in terms of computation requirement, storage requirement, and errors in the reconstructed signal.

1.4 Major Contributions of the Book

The major emphasis of this book can be delineated as below:

- Formulation of parallel filter structure algorithm of corresponding pyramid structure algorithm of DWT.
- Detailed analysis of filter coefficient quantization and round-off noise introduced due to fixed-point implementation of DWT algorithms.
- A mathematical model has been proposed to predict the signal-to-noise ratio (SNR in dB) of reconstructed signal in terms of number of bits, signal length, and depth of decomposition.
- Developed model has been applied to assess the suitability of parallel filter implementation of DWT algorithm using fixed-point arithmetic.
- Detailed study of signal denoising scheme based on wavelet thresholding has been made.
- Parallel implementation of DWT algorithm has been proposed on a cluster of distributed computing system in PVM environment in order to assess the speedup obtained.
- The denoising performance improvement on benchmark test images is presented.
- A novel method of classification, applying DWT for power quality problems, is proposed.
- The proposed scheme has been generalized with incorporation of soft computing technique (fuzzy rules).

Chapter 2
Filter Banks and DWT

Abstract The study of digital signal processing normally concentrates on the design, realization, and application of single-input, single-output digital filters. There are applications, as in the case of spectrum analyzer, where it is desired to separate a signal into a set of sub-band signals occupying, usually nonoverlapping, portions of the original frequency band. In other applications, it may be desired to combine many such sub-band signals into a single composite signal occupying the whole Nyquist range. To this end, digital filter banks play an important role. Implementation of a filter bank on a processor with finite precision arithmetic necessitates quantization of filter coefficients [95]. This results in loss of perfect reconstruction (PR) property. The theory of filter banks were developed much before modern discrete wavelet transform (DWT) analysis became popular [127, 134]. The study of literature reveals a close relationship between the DWT and digital filter banks. It turns out that a tree of digital filter banks, without computing mother wavelets, can simply achieve the wavelet transform. Hence, the filter banks have been playing a central role in the area of wavelet analysis. It is therefore of interest to study the filter bank theory before addressing the implementation issues of finite precision wavelet transforms. In this chapter, fundamental concept of filter bank theory leading to new implementation issues described in latter chapters is introduced. The material presented in this chapter will be useful in discussing error modeling and parallel computing techniques discussed in the book. In present chapter, the filter bank concept related to DWT is revisited in Sect. 2.1. Section 2.2 presents two-channel PR filter bank. Section 2.3 presents derivation of parallel filter DWT from pyramid DWT structure. Section 2.4 presents frequency response of generated parallel filters followed by conclusion in Sect. 2.5.

Keywords Filter banks · Quantum mirror filter · Aliasing · Computation complexity

K. K. Shukla and A. K. Tiwari, *Efficient Algorithms for Discrete Wavelet Transform*,   21
SpringerBriefs in Computer Science, DOI: 10.1007/978-1-4471-4941-5_2,
© K. K. Shukla 2013

2.1 Introduction

In the classical applications of multirate filter banks, a bank of analysis filters is applied to a discrete input signal and then down sampled at fixed rate to produce a set of sub-band signals. If a dual bank of synthesis filters exists, by means of which the original input signal can be recovered by first upsampling each of the above sub-band signals and then applying it to a synthesis filter, then the two filter banks are said to be a perfect reconstruction (PR) pair of filter banks [113]. The term uniform filter bank (UFB) is used to emphasize that all the sub-band signals are downsampled at the same rate [125]. PR pair of wavelet analysis and synthesis filter banks is dual. The discrete wavelet transform (DWT), and multiresolution analysis, can be viewed as the application of a *nonuniform filter bank*, defined by a UFB. In terms of wavelet theory, a *low-pass* filter corresponds *to scaling* function and the subsequent *high-pass* or *band-pass* filter corresponds to *wavelet* function. The DWT computation involves repetitive application of UFB on the *low-pass* channel. In the literatures, wavelet transform have been treated in considerable detail and wavelet decompositions have been related to PR Filter Bank [35, 132, 134, 139].

The concept of PR is meaningful only in the ideal cases. In most real world applications of finite world length, some sort of error is always introduced in the coding process or during the transmission over lossy channels. The advantage of multiresolution scheme is that the redundancy is introduced more in low-frequency channels compared to high-frequency channels. Thus, these representations may be advantageous for certain classes of signals such as natural images.

2.2 Orthogonal Filter Banks

The *digital filter bank* is defined as a set of digital band-pass filters with either a common input or a summed output and is referred as *analysis* and *synthesis* filter bank, respectively. The operation of *analysis* and *synthesis* filter bank is dual to each other. The combined structure of *analysis* and *synthesis* filter bank is quadrature mirror filter (QMF) bank [113].

Process of filtering is usually related with frequency selectivity. For example, an ideal discrete-time *low-pass* filter with cutoff frequency $\omega_c < \pi$ takes any input signal and projects it onto the subspace of signals bandlimited to $[-\omega_c, \omega_c]$. Orthogonal discrete-time filter banks perform a similar projection. Assume a discrete-time filter with finite impulse response $g_g[n] = \{g_g[0], g_g[1], \ldots, g_g[L]\}$, L even, and the property [107]

$$\langle g_g[n], g_g[n - 2k] \rangle = \delta_k \qquad (2.1)$$

that is, the impulse response is orthogonal to its even shifts, and $\left\|g_g\right\|_2 = 1$ [107]. The z-transform of impulse response $g_g[n]$ is

$$G_g[z] = \sum_{n=0}^{L-1} g_g[n]z^{-n}. \tag{2.2}$$

Further, with an assumption that $g_g[n]$ is a *low-pass* filter, corresponding *high-pass* filter $g_h[n]$ with z-transform, is given as follows:

$$G_h[z] = z^{-L+1}G_g(-z^{-1}). \tag{2.3}$$

Here, three operations have been applied [107] as follows:

1. $z \rightarrow -z$ corresponds to modulation by $(-1)^n$ or transforming the *low pass* into *high pass*.
2. $-z \rightarrow -z^{-1}$ applies time reversal to the impulse response.
3. Multiplication by z^{-L+1} makes the time-reversed impulse response causal.

This special way of obtaining a *high pass* from a *low pass*, introduced as quadrature conjugate filter (QCF) [128], has the following properties:

$$\langle g_h[n], g_h[n - 2k]\rangle = \delta_k \tag{2.4}$$

that is, the impulse response is orthogonal to its even shifts and

$$\langle g_g[n], g_h[n - 2k]\rangle = 0 \tag{2.5}$$

or the impulse response $\{g_g[n], g_h[n]\}$ and their even shifts are mutually orthogonal. Further, $\{g_g[n - 2k], g_h[n - 2l]\}_{k,l \in Z}$ is an orthonormal basis for $L_2(Z)$, the space of square summable sequences. Thus, any sequence from $L_2(Z)$ can be written as follows:

$$x[n] = \sum_{k \in Z} \alpha_k g_g[n - 2k] + \sum_{l \in Z} \beta_l g_h[n - 2l] \tag{2.6}$$

where $\alpha_k = \langle g_g[n - 2k], x[n]\rangle$ and $\beta_l = \langle g_h[n - 2l], x[n]\rangle$, k and $l \in Z$.

2.2.1 Two-Channel Quadrature Mirror Filter Bank

In filter bank applications, a discrete-time signal $x[n]$ is split into sub-band signals by means of an *analysis filter bank*. The sub-band signals are then processed and finally combined by a *synthesis filter bank* resulting in an output signal $y[n]$. If the sub-band signals are bandlimited to frequency ranges much smaller than that of the original input signal, they could be *downsampled* before processing. Due to lower sampling rate, the processing of the downsampled signals can be carried out more

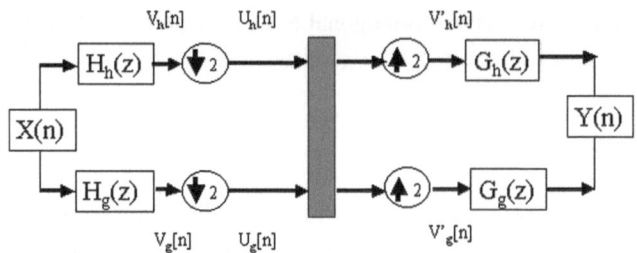

Fig. 2.1 Two-channel filter bank. $H_h(z)$ and $H_g(z)$ form an analysis filter bank, whereas $G_h(z)$ and $G_g(z)$ form a synthesis filter bank

efficiently. After processing, these signals are *upsampled* before being combined by the *synthesis filter bank* into a higher-rate signal. The filter bank theory dealt in detail in literature [80, 100, 127, 134, 139] is discussed in brief in this section.

Once the *low-pass* and *high-pass* filters have been computed, it is possible to compute the *scaling* function and the *mother wavelet*. Moreover, under certain conditions, the outputs of the *high-pass* filters are good approximations of the *wavelet* series. Consequently, the selection of desired *scaling* function and *mother wavelets* reduces to the design of *low-pass* and *high-pass* filters of two-channel PR filter banks. A tree of two-channel PR filter banks can simply realize the wavelet transform. Figure 2.1 sketches a typical two-channel PR filter bank system. It is convenient to analyze the filter bank in z-domain. As shown in Fig. 2.1, the signal $X(z)$ is first filtered by a filter bank consisting of $H_h(z)$ and $H_g(z)$.

The outputs of $H_h(z)$ and $H_g(z)$ are downsampled by 2 to obtain $U(z)$. After some processing, the modified signals are upsampled and filtered by another filter bank consisting of $G_h(z)$ and $G_g(z)$. The downsampling operators are decimators, and the upsampling operators are expanders. If no processing takes place between the two filter banks (in other words, $U(z)$ are not altered), the sum of the outputs of $G_h(z)$ and $G_g(z)$ is identical to the original signal $X(z)$, except for a time delay. Such a system is commonly referred to as a two-channel PR filter bank. $H_h(z)$ and $H_g(z)$ form an analysis filter bank, whereas $G_h(z)$ and $G_g(z)$ form a synthesis filter bank. The z-transform of input–output relations is defined as given by *Upil* in this chapter [26];

$$V_k(z) = H_k(z)X(z) \tag{2.7}$$

$$U_k(z) = \frac{1}{2}\left\{ V_k(z^{\frac{1}{2}}) + V_k(-z^{\frac{1}{2}}) \right\} \tag{2.8}$$

$$\hat{V}_k(z) = U_k(z^2) \tag{2.9}$$

where k refers to h and g (h and g are outputs of *high-pass* and *low-pass* filters, respectively).

Further, it can be shown that

$$
\begin{aligned}
\hat{V}_k(z) &= \frac{1}{2}\{V_k(z) + V_k(-z)\} \\
&= \frac{1}{2}\{H_k(z)X(z) + H_k(-z)X(-z)\}
\end{aligned}
\tag{2.10}
$$

and the reconstructed output of the filter bank is given by

$$
Y(z) = \frac{1}{2}\{G_h(z)\hat{V}_h(z) + G_g(z)\hat{V}_g(z)\}.
\tag{2.11}
$$

Substituting Eq. (2.10) in (2.11), the output of the filter bank is given as follows:

$$
\begin{aligned}
Y(z) &= \frac{1}{2}\{H_g(z)G_g(z) + H_h(z)G_h(z)\}X(z) \\
&+ \frac{1}{2}\{H_g(-z)G_g(z) + H_h(-z)G_h(z)\}X(-z)
\end{aligned}
\tag{2.12}
$$

The second term in the above equation is precisely due to aliasing caused by sampling rate alteration. The above equation is rewritten as follows:

$$
Y(z) = T(z)X(z) + A(z)X(-z)
\tag{2.13}
$$

where

$$
T(z) = \frac{1}{2}\{H_g(z)G_g(z) + H_h(z)G_h(z)\}
\tag{2.14}
$$

is called distortion transfer function and

$$
A(z) = \frac{1}{2}\{H_g(-z)G_g(z) + H_h(-z)G_h(z)\},
\tag{2.15}
$$

the term with $X(-z)$, is traditionally called the *aliasing term* matrix.

The relation for $Y(z)$ may be expressed in the matrix form as follows:

$$
Y(z) = \frac{1}{2}[X(z) \quad X(-z)]\begin{bmatrix} H_g(z) & H_h(z) \\ H_g(-z) & H_h(-z) \end{bmatrix}\begin{bmatrix} G_g(z) \\ G_h(z) \end{bmatrix}
\tag{2.16}
$$

The 2×2 matrix in the above equation is given as follows:

$$
H(z) = \begin{bmatrix} H_g(z) & H_h(z) \\ H_g(-z) & H_h(-z) \end{bmatrix}
\tag{2.17}
$$

In general, the QMF structure discussed above is a linear time-varying system. However, it is possible to select the analysis and synthesis filters such that the aliasing effect is canceled, resulting in a linear time-invariant (LTI) operation. To this end, we need to ensure that

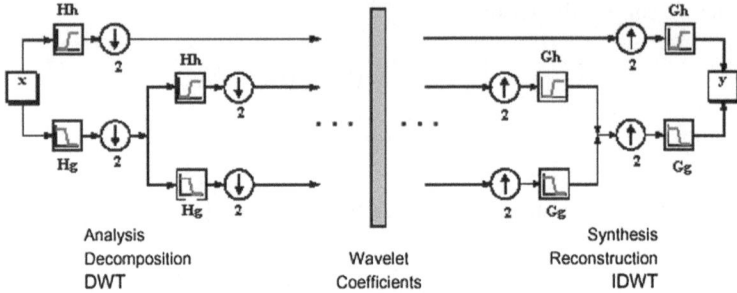

Analysis
Decomposition
DWT

Wavelet
Coefficients

Synthesis
Reconstruction
IDWT

Fig. 2.2 Multilevel wavelet decomposition/reconstruction using multiple PRQMF bank

$$2A(z) = \{H_g(-z)G_g(z) + H_h(-z)G_h(z)\} = 0 \qquad (2.18)$$

There are various possible solutions of the above equation. One solution may be given by

$$G_g(z) = H_g(-z), \quad G_h(z) = -H_g(-z). \qquad (2.19)$$

If above relation holds, then Eq. (2.13) reduces to

$$Y(z) = T(z)X(z) \qquad (2.20)$$

with

$$T(z) = \frac{1}{2}\{H_g(z)H_h(-z) - H_h(z)H_g(-z)\} \qquad (2.21)$$

Thus, an orthogonal filter bank splits the input space into *low-pass* approximation space V_g and its *high-pass* orthogonal component V_h. The space V_g corresponds to a *coarse approximation*, while V_h contains additional *details*. This is the first step in the multiresolution analysis that is obtained when iterating the *high-pass/low-pass* division on the *low-pass* branch (Fig. 2.1).

If an alias-free QMF bank has no amplitude and phase distortion, then it is called a perfect reconstruction mirror filter (PRQMF) bank. The time domain equivalent of the output is given by [100]

$$y(n) = dx(n - n_0) \qquad (2.22)$$

for all possible inputs. This indicates that the reconstructed output $y(n)$ is a scaled and delayed replica of the input. Thus, it is evident that the output resembles with the basic properties of the wavelet decomposition/reconstruction. The combination of multiple PRQMF bank results in multilevel wavelet decomposition/reconstruction as shown in Fig. 2.2.

2.2.2 Computational Complexity of Discrete Wavelet Transform

Rioul et al. in their seminal paper [106] have studied the computational complexity of wavelet transforms in detail. In general, the computations are periodic in 2^m for an m-level wavelet. Here, each filtered output is decimated by a factor of 2. This necessitates computation of those signal samples that are not thrown away. Consider an input set of $N = 2^m$ samples. For the first level, each filter computes $N/2$ samples, so the total number of samples generated at the *low-pass* and *high-pass* filters of level-1 wavelet is N. Similarly, each filter in the second-level wavelet computes $N/4$ samples, and the total number of samples computed at level 2 is $N/2$. In an m-level wavelet, the total number of samples computed is

$$N + \frac{N}{2} + \frac{N}{4} + \cdots\cdots + 2 = 2(N - 1). \tag{2.23}$$

Since the wavelet computation is periodic with N samples, the number of samples computed every sample period is $\frac{2(N-1)}{N}$ or $2\left(1 - \frac{1}{N}\right)$, which is upper bounded by 2 [106]. This implies that the maximum number of filters needed for computation in a one-dimensional multilevel forward wavelet transform is two. In other words, one *low-pass* and one *high-pass* filter will always be adequate for computation of one-dimensional DWT. The parallel filter bank structure discussed in next Section will lead to an efficient means for computation of wavelet transform.

2.3 Parallel Filter Bank Realization of Multilevel Discrete Wavelet Transform

As the computation of DWT involves filtering, an efficient filtering process is essential in DWT hardware implementation. A possible solution is based on Mallat algorithm [87] requiring only two filters (one *high-* and one *low*-pass filter). In the multistage DWT, coefficients are calculated recursively, and in addition to the wavelet decomposition stage, extra space is required to store the intermediate coefficients. Hence, the overall performance depends significantly on the precision of the intermediate DWT coefficients [74] as discussed in detail in next chapter. An alternative method for fast and efficient implementation of DWT transform is based on parallel filter implementation. In this, cascaded *high-pass* and *low-pass* filters at different resolution levels will be replaced by their equivalent filter [80], [107]. This necessitates number of filters to be of the order of decomposition level. The main advantage of the parallel filter algorithm is that it does not require storing intermediate coefficients [123]. Another advantage of this architecture is that the word length can be arbitrary and is not restricted to be a multiple of 2^m for m-resolution-level wavelet decomposition.

As discussed, Fig. 2.2 is a multilevel representation of DWT. The DWT evaluation is based on binary tree structured QMF. The output from *high-pass* filter is termed as detailed wavelet coefficients and from *low-pass* filter is termed as approximation coefficients. The approximation coefficients from previous level, after passing through another PRQMF filter bank, generate another set of detailed and approximation coefficients, and the decomposition process is continued until one reaches desired level of decomposition. The limitation here is that if the DWT coefficients of level L are of use, one has to first obtain the DWT coefficients at level $L - 1$, thus increasing computational burden. Souani et al. [123] presented an efficient one-dimensional direct DWT computation algorithm. The algorithm enables computation of Lth-level DWT coefficients without prior knowledge of $(L - 1)$th-level DWT coefficients. The algorithm is simple and uses a modified filter structure generated out of basic PRQMF filter bank. As discussed in next chapters, the algorithm is suitable from finite precision and parallel implementation viewpoint. Its implementation necessitates, finding equivalent parallel filters generated out of PRQMF filter bank to compute the DWT coefficient at any level from signal itself.

2.3.1 Iterated Filters and Regularity

The DWT filters roughly correspond to octave band filters. In many applications, *low-frequency* content of the signal is an important part. It is what gives the signal its identity. The *high-frequency* content, on the other hand, imparts flavor. For example, in the human voice, removing *high-frequency* components sounds different, but contents can still be inferred. However, removal of the *low-frequency* components sounds gibberish.

It is required to find the equivalent filter corresponding to the lower branch in Fig. 2.2 that is the iterated *low-pass* filter. It can be easily checked that subsampling by two followed by filtering with $G(z)$ is equivalent to filtering with $G(z^2)$ followed by the subsampling [80, 107]. Thus, the first two steps of *low-pass* filtering can be replaced with z-transform $G(z)$. $G(z^2)$, followed by subsampling by 4. In general, representing $G^J(z)$ the equivalent filter to the Jth stages of *low-pass* filtering and subsampling by 2^J [139]:

$$G^J(z) = \prod_{l=0}^{J-1} G(z^{2^l}) \tag{2.24}$$

A necessary condition for the iterated functions to converge to a continuous limit is that the filter $G(z)$ should have sufficient number of zeros at $z = -1$, or half sampling frequency, so as to attenuate repeat spectra [107]. Using this condition, the regular filters, which are both orthogonal and converge to continuous functions with compact support, may be generated. The well-known Daubechies orthonormal filters [36] are deduced from maximally flat *low-pass* filters.

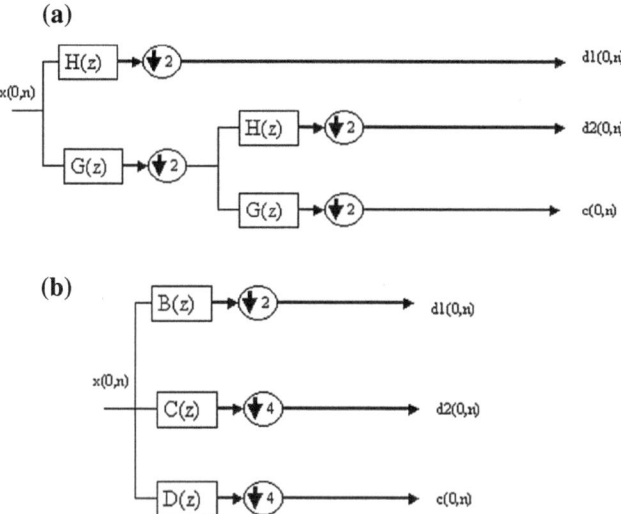

Fig. 2.3 Parallel filter implementation of two-level DWT decomposition. **a** Pyramid structure DWT. **b** Parallel filter DWT

2.3.1.1 Generation of Parallel Filter Banks

In present chapter for the sake of simplicity, the algorithm has been demonstrated only for two levels and three levels of DWT decomposition. For L level, it can be generalized there from. Consider the two-level DWT decomposition Mallat's algorithm [87] and derived parallel filter equivalent as shown in Fig. 2.3.

The equivalent analysis filters for two-level DWT (Fig. 2.3) are expressed in terms of PRQMF filter bank as follows:

$$B(z) = H(z) \tag{2.25}$$

$$C(z) = G(z)H(z^2) \tag{2.26}$$

$$D(z) = G(z)G(z^2) \tag{2.27}$$

Similarly, the equivalent analysis filters for three-level DWT, as shown in Fig. 2.4, are expressed in terms of PRQMF filter bank as follows:

$$B(z) = H(z) \tag{2.28}$$

$$C(z) = G(z)H(z^2) \tag{2.29}$$

$$D(z) = G(z)G(z^2)H(z^4) \tag{2.30}$$

$$E(z) = G(z)G(z^2)G(z^4) \tag{2.31}$$

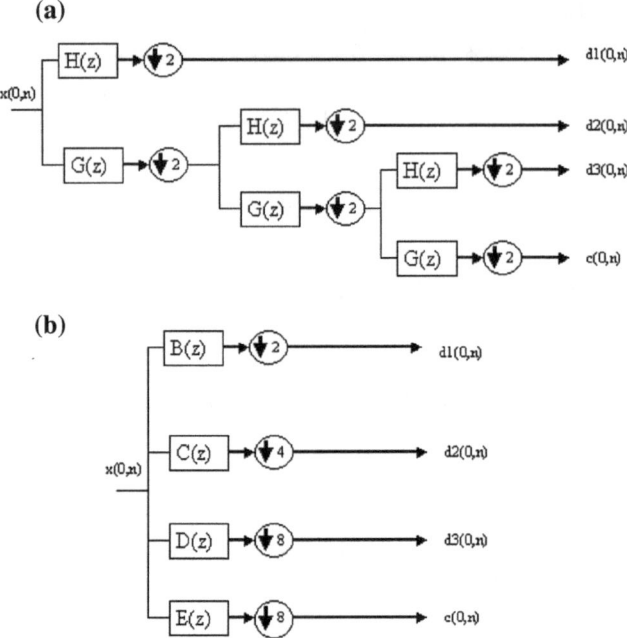

Fig. 2.4 Parallel filter implementation of three-level DWT decomposition. **a** Pyramid structure DWT. **b** Parallel filter DWT

More generally for J-level decomposition, the equivalent filter to J stages of low-pass filtering and subsampling by two (a total subsampling by 2^J) is given by [107]

$$E^J(z) = \prod_{l=0}^{J-1} G(Z^{2^l}) \tag{2.31}$$

This explains that the generated filter length will increase with increase in depth of decomposition. The equivalent synthesis filters can be generated accordingly.

2.3.2 One Set of Forward Discrete Wavelet Transform Computation

The computation of the 3-level wavelet is periodic with period 2^3 (or 8), that is, identical sets of computations are separated by a time index of 2^3 [4]. To explain the generated parallel filter bank structure, it is required to write down the set of computations associated with one period in forward DWT decomposition. This set completely describes the computations. All other computations are generated from

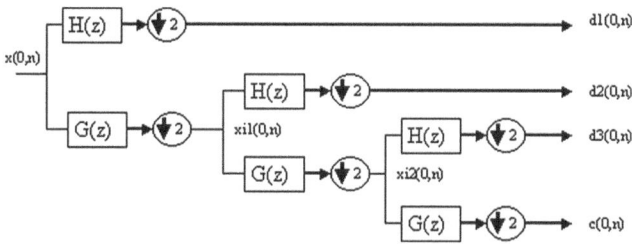

Fig. 2.5 Implementation of three-level DWT decomposition and intermediate coefficients ($xi1$ and $xi2$ are input to PRQMF at level two and three, respectively)

this set by shifting the time by multiples of the period. For simplicity, following transfer function representation of filters used in PRQMF filter bank with L–tap filters is assumed as follows:

$$H(z) = \sum_{n=0}^{L-1} h(z)z^{-n} \tag{2.32}$$

$$G(z) = \sum_{n=0}^{L-1} g(z)z^{-n} \tag{2.33}$$

For simplicity, the filter tap is selected to $L = 6$.

For details of multilevel DWT coefficient computation, readers are advised to see [60, 111, 123]. One period of DWT computation, described in Fig. 2.5, is as follows:

The first-level computations are as follows:

$$
\begin{aligned}
d1(0) &= h(0)x(0) + h(1)x(-1) + h(2)x(-2) + h(3)x(-3) + h(4)x(-4) + h(5)x(-5)\\
d1(2) &= h(0)x(2) + h(1)x(1) + h(2)x(0) + h(3)x(-1) + h(4)x(-2) + h(5)x(-3)\\
d1(4) &= h(0)x(4) + h(1)x(3) + h(2)x(2) + h(3)x(1) + h(4)x(0) + h(5)x(-1)\\
d1(6) &= h(0)x(6) + h(1)x(5) + h(2)x(4) + h(3)x(3) + h(4)x(2) + h(5)x(1)\\
xi1(0) &= g(0)x(0) + g(1)x(-1) + g(2)x(-2) + g(3)x(-3) + g(4)x(-4) + g(5)x(-5)\\
xi1(2) &= g(0)x(2) + g(1)x(1) + g(2)x(0) + g(3)x(-1) + g(4)x(-2) + g(5)x(-3)\\
xi1(4) &= g(0)x(4) + g(1)x(3) + g(2)x(2) + g(3)x(1) + g(4)x(0) + g(5)x(-1)\\
xi1(6) &= g(0)x(6) + g(1)x(5) + g(2)x(4) + g(3)x(3) + g(4)x(2) + g(5)x(1)
\end{aligned}
$$

$$\tag{2.34}$$

The second-level computations are as follows:

$$
\begin{aligned}
d2(0) &= h(0)xi1(0) + h(1)xi1(-2) + h(2)xi1(-4) + h(3)xi1(-6) + h(4)xi1(-8) + h(5)xi1(-10)\\
d2(4) &= h(0)xi1(4) + h(1)xi1(2) + h(2)xi1(0) + h(3)xi1(-2) + h(4)xi1(-4) + h(5)xi1(-6)\\
xi2(0) &= g(0)xi1(0) + g(1)xi1(-2) + g(2)xi1(-4) + g(3)xi1(-6) + g(4)xi1(-8) + g(5)xi1(-10)\\
xi2(4) &= g(0)xi1(4) + g(1)xi1(2) + g(2)xi1(0) + g(3)xi1(-2) + g(4)xi1(-4) + g(5)xi1(-6)
\end{aligned}
$$

$$\tag{2.35}$$

The third-level computations are as follows:

$$d3(0) = h(0)xi2(0) + h(1)xi2(-4) + h(2)xi1(-8) + h(3)xi1(-12) + h(4)xi2(-16) + h(5)xi2(-20)$$
$$c(0) = g(0)xi2(0) + g(1)xi2(-4) + g(2)xi2(-8) + g(3)xi2(-12) + g(4)xi2(-16) + g(5)xi2(-20)$$
$$(2.36)$$

The variables $d1$, $d2$, $d3$, c, x, $xi1$, and $xi2$ are appropriately defined in Fig. 2.5. The negative time indexes in these equations correspond to the reference starting time unit 0. By adding one or multiples of the periods of computation to these equations, the next sets of computations are obtained. The condensed form of Eqs. (2.34–2.36) [123] is as follows:

$$d1(2k) = \sum_{p=0}^{L-1} h(p)x(2k - p) \tag{2.37}$$

$$xi1(2k) = \sum_{p=0}^{L-1} g(p)x(2k - p) \tag{2.38}$$

$$d2(4k) = \sum_{p=0}^{L-1} h(p)xi1(4k - 2p) \tag{2.39}$$

$$xi2(4k) = \sum_{p=0}^{L-1} g(p)xi1(4k - 2p) \tag{2.40}$$

$$d3(8k) = \sum_{p=0}^{L-1} h(p)xi2(8k - 4p) \tag{2.41}$$

$$c(8k) = \sum_{p=0}^{L-1} g(p)xi2(8k - 4p) \tag{2.42}$$

In the above equations, the coefficients obtained by $d1$, $d2$, $d3$, and c are final DWT coefficients and coefficients obtained by $xi1$ and $xi2$ are intermediate. Variable x denotes the input signal. The DWT computation is complex because of the data dependencies at different octaves. Above equations show the relationship among final and intermediate coefficients.

Implementation of the three-level DWT necessitates total of six filters to be used. The filters are a pair of identical PRQMF filter bank used at each stage. The DWT coefficients could be derived in terms of input signal $x(n)$ only, thus eliminating the intermediate-level coefficients. This will lead derivation of new filters, as per Eqs. (2.28–2.31), enabling computation of DWT coefficients independent of intermediate coefficients. The filter B is the same as high-pass filter H with length $L^B = L$. The filter lengths of generated filters C, D, and E are as follows:

Table 2.1 Generated filters length in terms of base PRQMF filter length

PRQMF filter tap length	Generated filter length			
L	B $(L^B = L)$	C $(L^C = 3L - 2)$	D $(L^D = 7L - 6)$	E $(L^E = 7L - 6)$
4	4	10	22	22
6	6	16	36	36
8	8	22	50	50
10	10	28	64	64
12	12	34	78	78

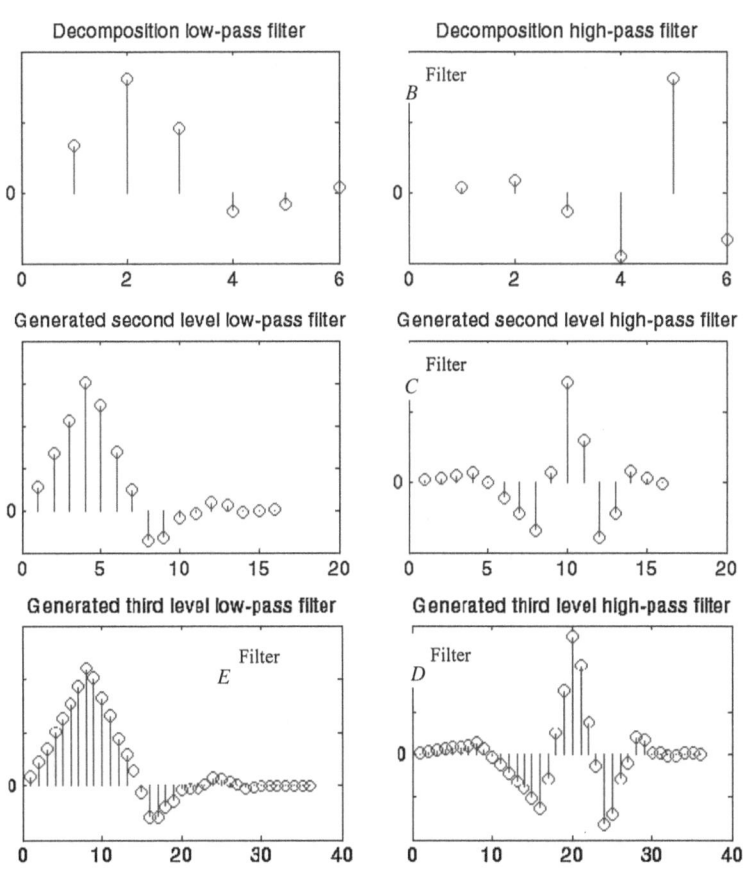

Fig. 2.6 Impulse response plot of generated parallel filters for three-level DWT, (x-axis: filter coefficient number and y-axis: corresponding magnitude)

$$L^C = 3L - 2$$
$$L^D = 7L - 6 \qquad (2.43)$$
$$L^E = 7L - 6$$

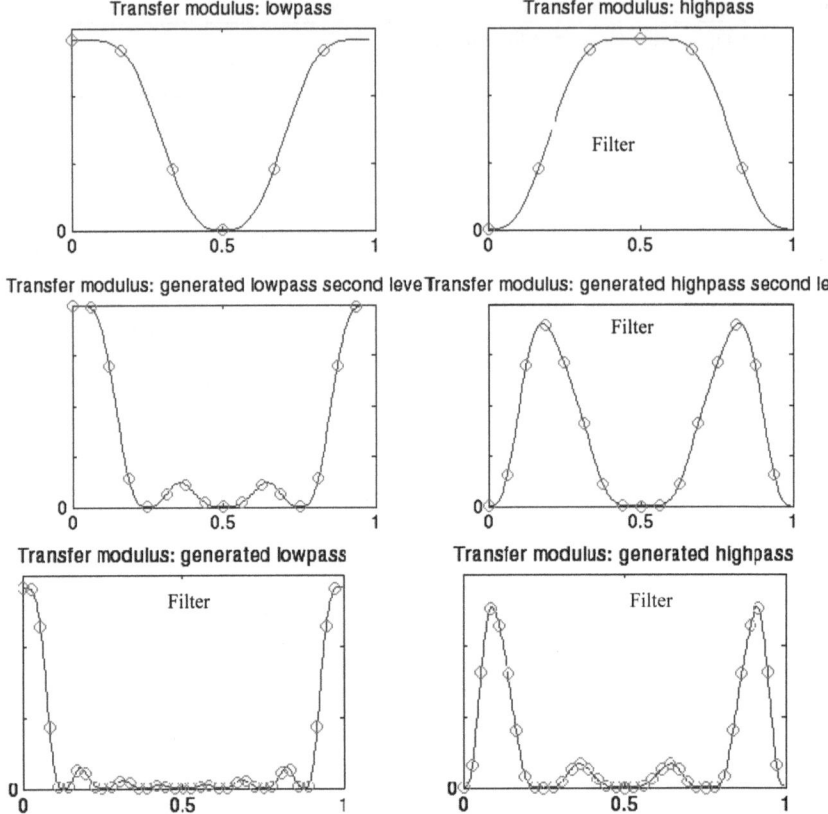

Fig. 2.7 Frequency response of generated parallel filters for three-level DWT

The generated parallel filters lengths for varied PRQMF filter lengths are given in Table 2.1. It is evident that the filter B operates every two samples (downsampling by 2). Filter C operates every four samples; filters D and E operate every eight samples. For an even order of the input data, filters B, C, D, and E will operate depending on their decimation rate.

2.4 Frequency Response of Generated Parallel Filter Bank

To validate the parallel filter DWT structure, frequency response plots are generated. The frequency response plots corresponding to three-level DWT decomposition are shown. The selected PRQMF filter bank is a Daubechies filter [37] with six taps, and Symlet filter [14] with eight taps. The experimentation has been carried out on a Pentium III, 733 MHz system using Matlab [93].

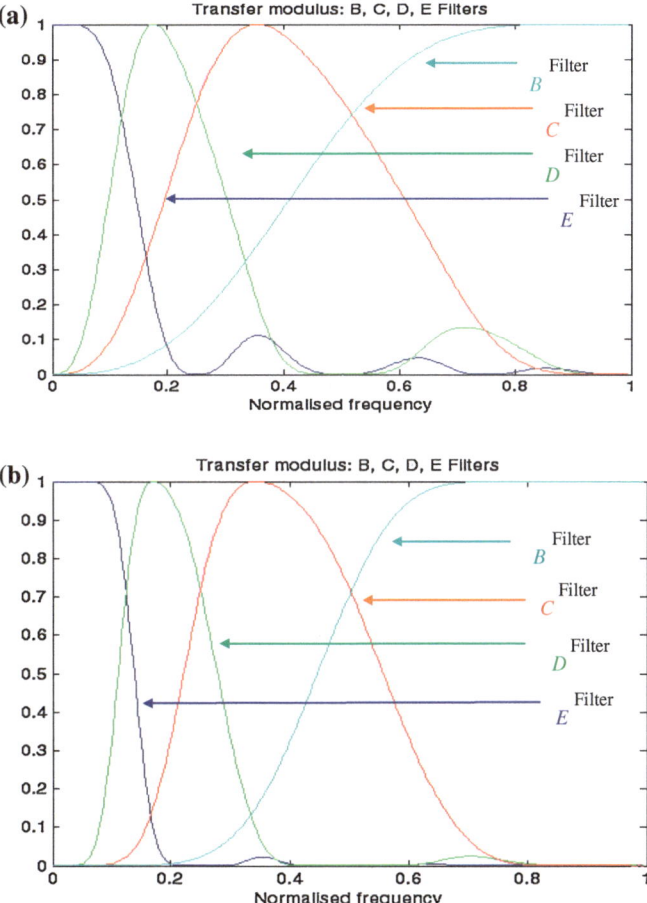

Fig. 2.8 **a** Frequency plot of generated parallel filter structure (Daub 6-tap PRQMF filter bank), **b** Frequency plot of generated parallel filter structure (Symlet 8-tap PRQMF filter bank)

Figure 2.6 is an impulse response plot of generated parallel filter structure. Figure 2.7 plots frequency response of generated parallel filter for two- and three-level DWT decomposition. It is evident from the plots that for one set of PRQMF filter bank, the generated parallel filters do confirm the frequency response desired at various levels. The parallel filter corresponding to approximate DWT coefficients (filter E, Figs. 2.3b and 2.4b) resembles *low-pass* filter (*scaling* function). Filters C and D correspond to *band-pass* (*high-pass*) filters (*wavelet* function) and filter B corresponds to *high-pass* filter, which is in turn *high-pass* filter of PRQMF filter bank (Fig. 2.8). Figure 2.9 is gain plot of derived parallel filters. This again confirms suitability of parallel filter structure for DWT decomposition.

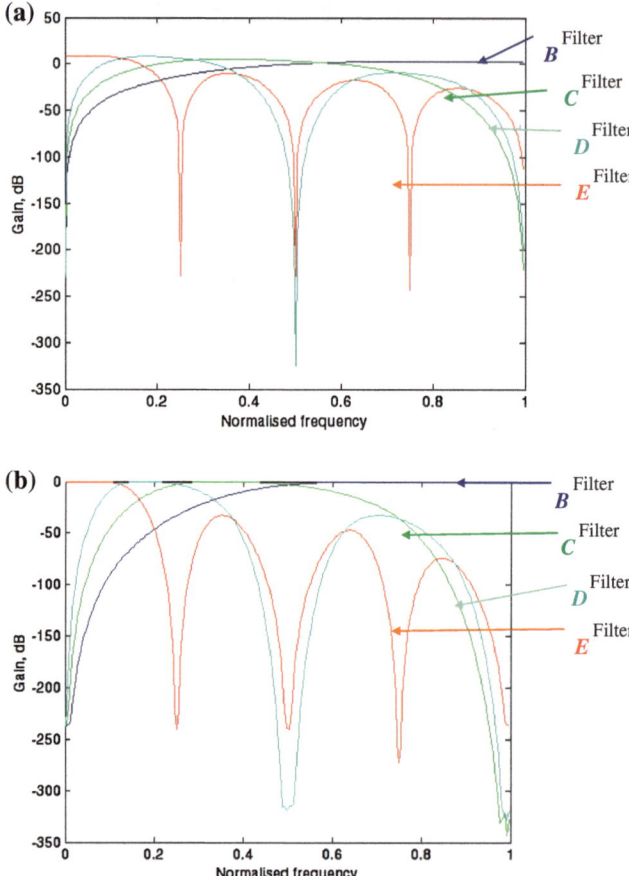

Fig. 2.9 a Gain plot of generated parallel filters for three-level DWT (db6), **b** Gain plot of generated parallel filters for three-level DWT (sym8 PRQMF)

2.5 Conclusions

The filter bank structure of DWT is analyzed. The relation for PR is presented. Computational complexity for DWT presented in this chapter will be basis for development of error analysis model in the next chapter. An alternative structure of DWT in terms of parallel filters is also derived. Impulse response and frequency response plots of generated parallel filter structure validate its suitability in terms of dyadic frequency selectivity. Chapter 3 presents comparison of finite precision error of the two models.

Chapter 3
Finite Precision Error Modeling and Analysis

Abstract Discrete wavelet transforms (DWTs) have excellent energy compaction characteristics and are able to provide near perfect reconstruction (PR). They are ideal for signal/image analysis and encoding. Hardware implementation of DWT is fast and area efficient in fixed-point arithmetic. DWT encoding has been drawing much attention because of its ability to decompose signals into a hierarchical structure that is suitable for adaptive processing in the transform domain. In existing architectural designs for the DWT, little consideration has been given to word size and precision. Present chapter addresses this problem, showing how the word size requirements can be calculated for a specific problem (based on the range of input data and wavelet used). A simplified, statistical model is proposed. As the depth of the DWT filtering increases, the data word length requirement increases. It is important to investigate how this can affect the potential of the resulting hardware implementation of DWT. The issue has been analyzed for both pyramid structure DWT and parallel filter DWT. The organization of this chapter is as follows. Section 3.1 presents background material related to subject. Section 3.2 presents in brief the computational complexity of DWT. Section 3.3 presents finite precision modeling of two-channel PR filter bank in moderate detail, including modeling of quantized coefficient filters. Section 3.4 presents the proposed statistical modeling of DWT to study the effects of finite word length implementation. This includes construction of new DWT filters to accommodate round-off errors followed by corresponding mathematical derivation.

Keywords Error modeling · Round-off noise · DWT

K. K. Shukla and A. K. Tiwari, *Efficient Algorithms for Discrete Wavelet Transform*, 37
SpringerBriefs in Computer Science, DOI: 10.1007/978-1-4471-4941-5_3,
© K. K. Shukla 2013

3.1 Introduction

Many applications of DWT in signal processing require high precision outputs. However, if sufficient precision is not available in computation, the perfect reconstruction (PR) is not necessarily possible due to rounding errors. The *decimation* and *interpolation* operation in sub-band filter bank with fixed filter coefficients results in a overall system that is generally linear but periodically time varying. Mostly the studies related to DWT have been with the coefficients of digital filters and resultant wavelet coefficients incorporating infinite precision (taking any value between $-\infty$ and $+\infty$) [4, 18, 23, 60, 106, 107, 132, 140, 141]. These structures guarantee PR of the original signal from its sub-band components in the absence of quantization and transmission errors. This means aliasing errors in sub-band analysis parts get canceled with that in synthesis part. However, when implemented in either software form on a general-purpose computer or special-purpose digital hardware, the system parameters along with the signal variables can take only discrete values, within a specified range, as the registers of the digital machines, where the results are stored, are of finite word length [108]. Gadre et al. [53] in their paper presented results on multirate filters with effects of coefficients perturbation due to finite word length. The authors discussed the issue by presenting a theorem for exact analysis. Implementation of digital signal processing algorithms on a digital computer or with special-purpose hardware introduces quantization errors: *input quantization, coefficient rounding,* and *arithmetic quantization* due to finite register length effects. Cancelation of these errors is not possible in the reconstruction stage. Characteristics of these errors are required to be known for realization of an algorithm with minimum word length and acceptable accuracy. It is often useful to perform an approximate analysis by representing the effect as an additive error signal, which will be referred to as round-off noise [76, 110]. Fortunately, if the quantization amounts are small compared to the values of the signal variables and filter constants, a simpler approximate theory based on a statistical model will be of use and it is possible to derive the effects of discretization and develop results that can be verified experimentally.

For a software implementation where speed may be of no concern, the programmer may be capable of supporting sufficient and/or increasing precision to accurately represent the values being computed, but for a hardware implementation, the precision is fixed at design time so must be accurately determined before building the hardware. In implementation of DWT algorithm either with special-purpose digital hardware or as programs, sequence values and coefficients are stored in a binary format with finite length registers. The finite word length constraint is manifested in a variety of ways.

A lot of research work has been done on quantization error of various algorithms with different implementations. Fixed-point and floating-point error analyses for FFT algorithms are well known [83, 142]. The coefficient round-off error in the FFT implementation with a logarithmic number system is presented by Oppenheim and Schafer [108, 109] in detail.

The focus of this chapter is the fixed-point effects that result from implementing an algorithm in hardware. In particular, fixed-point effects in implementation of DWT are studied. When an algorithm is implemented using fixed-point arithmetic, the quality of results can be significantly different from those obtained using floating-point arithmetic. In fixed-point arithmetic, the range of allowable data (including input data, intermediate results, and final output) is limited to a pre-specified number N_i of integer bits and N_f of fractional bits. In algorithms where the dynamic range of data is large, as in the case of an orthogonal wavelet transform, fixed-point effects can severely limit the quality of results obtained.

3.2 Computational Complexity of DWT

The DWT has been recognized as a natural wavelet transform for discrete-time signals by several authors [4]. As far as the structure of computation is concerned, the DWT is same as an octave-band filter bank. The filter bank has a regular structure; it is easily implemented by repeated application of identical cells Fig. 3.1 [106]. It is computationally efficient. The decomposition of real-time

Fig. 3.1 Basic computational cell of (**a**) the DWT, (**b**) the inverse DWT

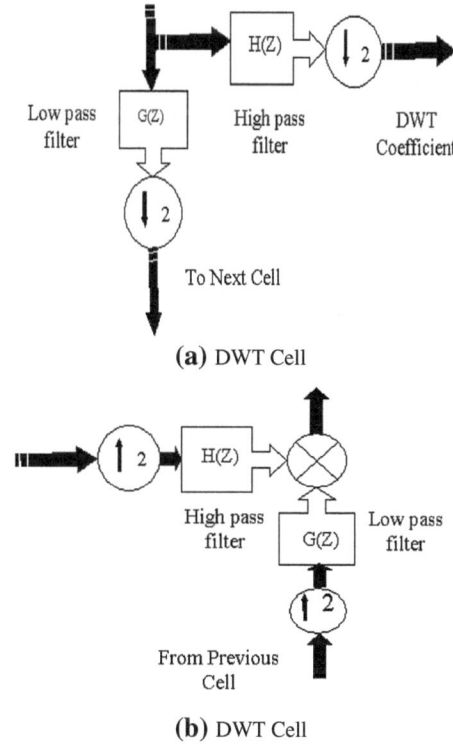

signal into wavelet coefficients involves FIR filtering and its reconstruction involves IIR filtering.

The wavelet filtering by *high pass* filter ($h[n]$) and *decimation* provides the wavelet coefficients at the considered octave. The filtering by *low pass* filter ($g[n]$) and decimating is used to enter the next cell. Thus, direct implementation of filters $h[n]$ and $g[n]$ followed by decimation requires $2L$ multiplication and $2(L\text{-}1)$ additions for every set of two inputs with L tap filters [106]. Thus, complexity per input for each elementary cell is

$$L \text{ mults/point/cell and } L - 1 \text{ adds/point/cell} \tag{3.1}$$

Since the cell at the Jth octave has input subsampled by 2^{J-1}, the total complexity required by a filter bank implementation of the DWT on J octaves is [106] $1 + \frac{1}{2} + \frac{1}{4} + \cdots + \frac{1}{2^{J-1}} = 2\left(1 - 2^{-J}\right)$ times the complexity in Eq. (3.1).

Thus, complexity per input at the end of Jth octave is

$$2L(1 - 2^{-J}) \text{ mults/point and } 2(L - 1)(1 - 2^{-J}) \text{ adds/point.} \tag{3.2}$$

The DWT is therefore roughly equivalent to, in terms of complexity, one filter length $2L$. The complexity remains bounded as the number of octaves J increases.

3.2.1 The DWT Data: Word Requirements

The finite register word length implementation of DWT introduces errors mainly due to accumulation of round-off errors in each multiplication step. Here, the effect of rounding or truncation is represented as an additive noise signal [108]. They are uncorrelated to each other and with the input signals. In the hardware implementation, fixed-point arithmetic finds wide application. In fixed-point implementation, round-off errors are introduced with each multiplication. The additions are free of errors provided that no overflow occurs.

3.3 Finite Precision Filter Bank Implementation

Decimation and interpolation in the sub-band filter bank with fixed filter coefficients result in an overall system that is generally linear but periodically time varying [26]. Conventional design techniques for *analysis* and *synthesis* filters guarantee PR of the original signal from its sub-band components in the absence of quantization and transmission errors. This assumes the aliasing errors in sub-band analysis part are explicitly canceled out in the reconstruction stage. However, in actual VLSI, implementation of DWT following quantization errors is present:

(1) *Input signal quantization error.*
(2) *Filter coefficient quantization (FCQ).*
(3) *Finite word length data path.*

This necessitates investigating the effect of FCQ in detail for implementation of DWT.

3.3.1 Coefficient Quantization in FIR Filters

Gadre et al. [53] in their seminal paper have presented an equivalence theorem to study the coefficient perturbations in implementation when finite world lengths are used for coefficients. Digital samples generated by A/D converter are binary representation of quantized version of an ideal sampler with infinite precision. As the output registers are of finite word length, the digital equivalent can take a value from a finite set of discrete values within the dynamic range of the register. Without loss of generality, we consider fixed-point numbers represented as $(b + 1)$ bit binary fractions including the sign bit. The binary point is just to the right of the highest order bit [108]. The total number of discrete levels available for representation is 2^{b+1}. We will also assume that the round-off error in multiplying two fixed-point b bit numbers has a uniform probability density function in the interval $\left(-\frac{1}{2}2^{-b}, +\frac{1}{2}2^{-b}\right)$, with variance $\sigma_e^2 = \frac{1}{12}2^{-2b}$ [108]. Furthermore, the round-off errors are assumed to be uncorrelated to each other and with the input. Based on these assumptions, the round-off noise is modeled by inserting additive independent signals into the flow graph and analyzing the effects of the noise sources on the output.

The direct form of realization of a linear shift-invariant system with unit sample response h[n], which is nonzero only for $0 \leq n \leq (N - 1)$, is a direct realization of convolution sum relation

$$y(n) = \sum_{k=0}^{N-1} h(k)x(n - k) \tag{3.3}$$

The effect of rounding the product is manifested in form of a new transfer function corresponding to FIR transfer function as:

$$\hat{H}(z) = H(z) + E(z) \tag{3.4}$$

where for an $(M - 1)$th order FIR filter $H(z)$ and $E(z)$ is given by

$$H(z) = \sum_{n=0}^{M-1} h[n]z^{-n} \tag{3.5}$$

Fig. 3.2 Model of the FIR
filter with quantized
coefficients

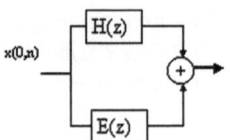

$$E(z) = \sum_{n=0}^{M-1} e[n]z^{-n} \tag{3.6}$$

Thus, the system function (and therefore, also the frequency response) of the quantized system is linearly related to the quantization errors in the impulse response coefficients [100]. Thus, the FIR filter with quantized coefficients can be modeled as a parallel connection of two filters, $H(z)$ and $E(z)$ as shown in Fig. 3.2.

In filter bank system, FCQ can result in deeper consequences. For example a QMF bank loses the alias free or PR property due to multiplier quantization.

Figures 3.3, 3.4 demonstrates effect of coefficient quantization on pole zero and impulse response plot of a FIR filter (order 12) respectively. Selected quantizer widths are 6 and 12 bits. It is clear from the plots that quantizing the filter coefficients with 6-bit, has seriously degraded its response. With 12 bit quantizer the filter performance is very much close to that of original filter. Thus selection of quantizer width plays a significant role in filters implementation.

3.3.2 Round-off Noise Model

The implementation of DWT consists of filtering and data routing. Inserting a quantizer after each FIR filter of the filter bank, results rounded outputs of the filters. In order to develop necessary mathematical model to analyze the finite world length effect following assumptions [108] are made with respect to output error sources:

1. The error sources $e_k(n)$ are white noise sources.
2. The round-off noise due to each real multiplication is uniformly distributed in amplitude between $\pm\frac{1}{2}2^{-b}$ and has a variance of $\sigma_e^2 = A\frac{1}{12}2^{-2b}$.
3. The errors are uniformly distributed over one quantization interval.
4. The error sources are uncorrelated with the input and with each other.

A constant gain, A, is applied which depends on the statistics of input signal and allowed overflow probability. Each noise source adds directly to the output, and therefore the output noise is

$$f(n) = \sum_{k=0}^{N-1} e_k(n) \tag{3.7}$$

Because the noise sources are assumed independent, the variance of the output noise (for rounding) is [108]:

Fig. 3.3 **a** Effect of coefficients quantization (6 bit) on pole zero plot of FIR filter (12 order), **b** effect of coefficients quantization (12 bit) on pole zero plot of FIR filter (12 order)

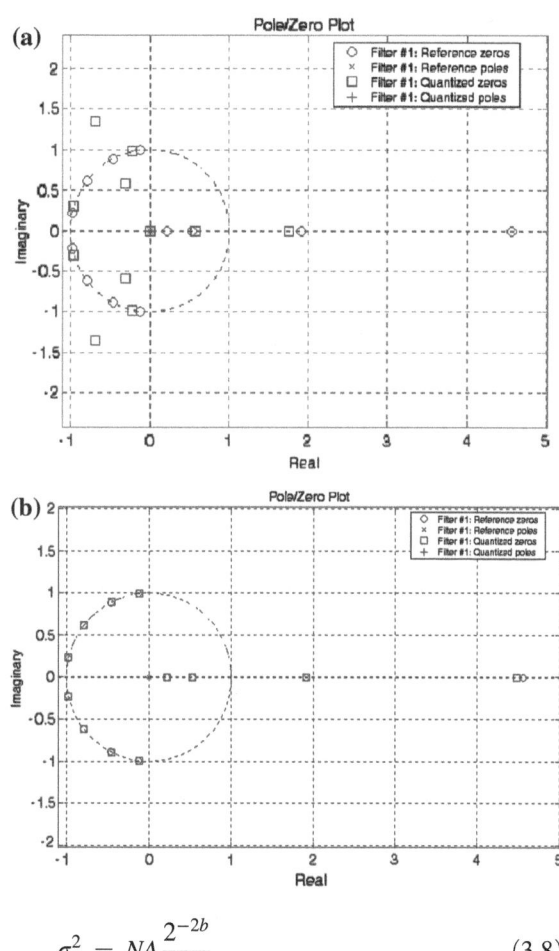

$$\sigma_f^2 = NA\frac{2^{-2b}}{12} \tag{3.8}$$

where N is length of input data and b is number of bits in computation.

The dynamic range limitation of fixed-point arithmetic necessitates scaling of the input so that no overflow occurs. A least upper bound on the output of linear shift-invariant system is

$$|y(n)| \leq x_{\max} \sum_{n=-\infty}^{\infty} |h(n)| \tag{3.9}$$

where $x_{\max}$ is the maximum magnitude of the input signal. In order to guarantee no overflow, the gain at the input should satisfy [60]

$$A \langle \frac{1}{x_{\max} \displaystyle\sum_{n=-\infty}^{\infty} |h(n)|} \tag{3.10}$$

Fig. 3.4 a Effect of
coefficients quantization (6
bit) on impulse response plot
of FIR filter (12 order),
b effect of coefficients
quantization (12 bit) on
impulse response plot of FIR
filter (12 order)

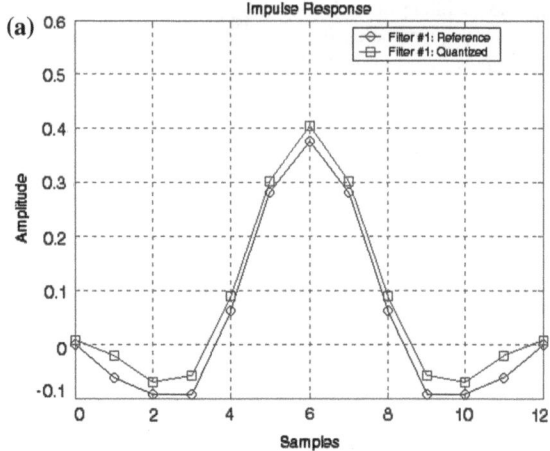

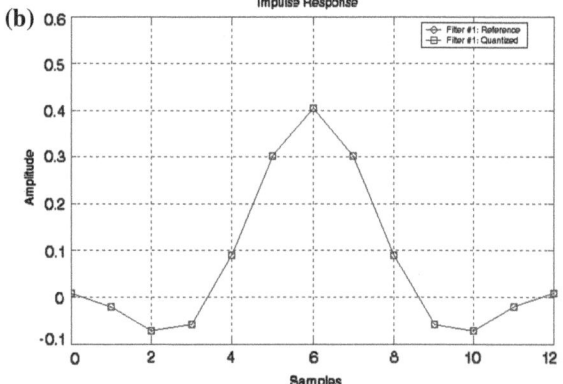

3.3.3 Quantized Coefficient Modeling of Perfect Reconstruction Filter Bank

Finite word length implementation of a QMF bank leads FCQ and round-off errors
produced by both the analysis and the synthesis filter bank [26, 132]. In addition
noise from analysis stage propagates to synthesis stage. The filter bank implementation of DWT as discussed in Chap. 2 without any quantization Eq. (3.13) is:

$$Y(z) = T(z)X(z) + A(z)X(-z) \qquad (3.11)$$

where $T(z) = \frac{1}{2}\{H_g(z)G_g(z) + H_h(z)G_h(z)\}$ (3.12) and $A(z) = \frac{1}{2}\{H_g(-z)G_g(z) + H_h(-z)G_h(z)\}$ (3.13)

With incorporation of quantized filter model we have

$$\hat{H}_g(z) = H_g(z) + E_{\text{Hg}}(z)$$

$$\hat{G}_g(z) = G_g(z) + E_{\text{Gg}}(z)$$

$$\hat{H}_h(z) = H_h(z) + E_{\text{Hh}}(z)$$

$$\hat{G}_h(z) = G_h(z) + E_{\text{Gh}}(z)$$

and resultant output

$$\hat{Y}(z) = \hat{T}(z)X(z) + \hat{A}(z)X(-z) \tag{3.14}$$

where

$$\hat{T}(z) = T(z) + E_T(z)$$

$$\hat{A}(z) = A(z) + E_A(z) \tag{3.15}$$

Thus the error signal due to FCQ at the output of QMF bank is given by

$$E(z) = Y(z) - \hat{Y}(z)$$
$$= \frac{1}{2}\left(E_T(z)X(z) + E_A(z)X(-z)\right) \tag{3.16}$$

Thus the error is composed of input signal quantization, round-off noise due to FCQ and round-off noise due to aliasing error. Further, the design of PR filters are such that it completely removes the aliased signal $X(-z)$ from the output. After eliminating aliasing term the transfer function of filter bank is given by $T(z)$. Thus final error is composed of *input signal quantization*, *round-off noise due to FCQ* and round-off noise due to *nontrivial multiplications per output*.

3.4 Finite Precision DWT Modeling

The accuracy of DWT coefficients depends on the precision of both the input data and the DWT coefficients. In this section, the performance of a finite precision DWT is evaluated. We note that a multistage DWT is calculated recursively by FIR filtering. Therefore, in addition to the wavelet decomposition stage, extra space is required to store the intermediate coefficients. Hence, the overall performance depends significantly on the precision of the intermediate DWT coefficients. It is assumed that both the intermediate and final DWT coefficients are represented with the same precision and their dynamic range is symmetric about zero.

Let us assume that the precisions of the various data are as follows:

(1) *Input data: i bits.*
(2) *Intermediate wavelet coefficients: j bits.*
(3) *Wavelet filter coefficients: m bits.*

Generally, the dynamic range of DWT coefficients is greater than the dynamic range of input data, and hence j should be greater than i. In our study, we multiply the input data by 2^{j-i}, to make them j bits precision. Subsequent stages of DWT decomposition are executed in the same manner.

To execute implementation of DWT, a multiplier of $j \times m$ bits is required. The dynamic range of the DWT coefficients will increase because of sum of several terms. The rate of sum will be upper bounded by $\sum |h(n)|$. Applying Eq. (3.9) recursively, one can deduce the maximum dynamic range of DWT coefficients at various stages as [60]

$$\text{Input data} \quad (-x_{\max}, x_{\max}) \tag{3.17a}$$

First-level Coefficients

$$\left(-x_{\max} \left(\sum |h(n)| \right), x_{\max} \left(\sum |h(n)| \right) \right) \tag{3.17b}$$

Second-level Coefficients

$$\left(-x_{\max} \left(\sum |h(n)| \right)^2, x_{\max} \left(\sum |h(n)| \right)^2 \right) \tag{3.17c}$$

Third-level Coefficients

$$\left(-x_{\max} \left(\sum |h(n)| \right)^3, x_{\max} \left(\sum |h(n)| \right)^3 \right) \tag{3.17d}$$

Normally, it has been observed that for a number of wavelet functions, the value of $\sum |h(n)|$ is less than two. Hence, for a one-dimensional discrete wavelet transform, the dynamic range will increase at most by two.

It is to be noted that there is a difference between the forward and inverse DWT coefficients. The dynamic range of the forward wavelet coefficients increases with the tree depth because of the summation operation. However, in the case of inverse DWT, the dynamic range of the coefficients will decrease due to the PR property of wavelet transform Fig. 3.5.

When all the stages have been computed, the resulting output data are with j bit precision. To obtain the original data, the output data should be divided by 2^{j-m}. Similarly, for a data sequence of $N = 2^m$ samples, for the first level, each filter

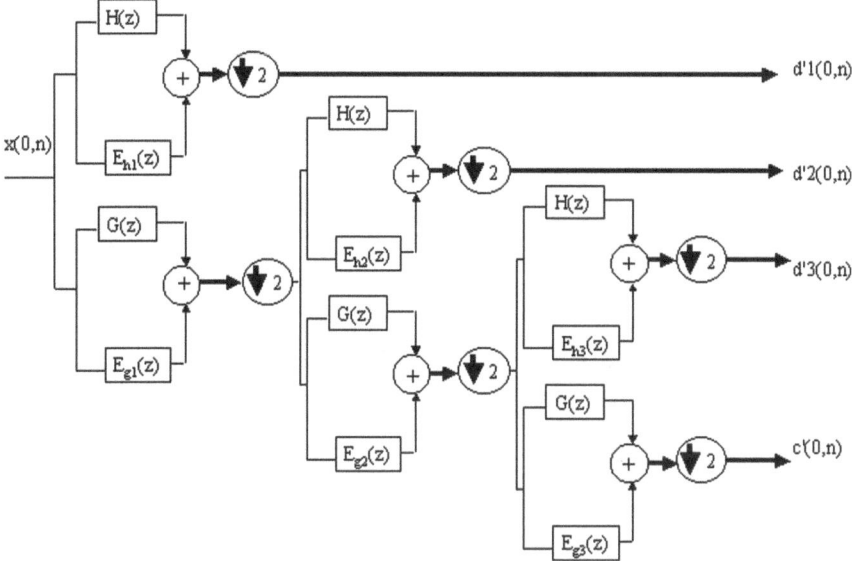

Fig. 3.5 Model for three-level DWT analysis filter bank with quantized coefficients

computes N/2 samples, with N/2 real multiplications. Hence, the variance of the output noise in first level detail coefficients will be

$$\sigma_{d1}^2 = \frac{NL}{2}\frac{2^{-2b}}{12} \tag{3.18}$$

Similarly, in the second-level computation, it takes N/4 multiplications and the total round-off error in coefficients at level-2 is

$$\sigma_{d2}^2 = \frac{NL}{2}\frac{2^{-2b}}{12} + \frac{NL}{4}\frac{2^{-2b}}{12} \tag{3.19}$$

Since the noise generated in each multiplier is assumed to be independent, the output error signal variance at mth level detailed wavelet coefficient is given by

$$\begin{aligned}
\sigma_{dm}^2 &= \frac{NL}{2^1}\frac{2^{-2b}}{12} + \frac{NL}{2^2}\frac{2^{-2b}}{12} + \frac{NL}{2^3}\frac{2^{-2b}}{12} + \cdots + \frac{NL}{2^m}\frac{2^{-2b}}{12} \\
&= NL(1 - 2^{-m})\frac{2^{-2b}}{12} \\
&= NL(1 - 2^{-m})\sigma_e^2
\end{aligned} \tag{3.20}$$

and will be same for approximate coefficients, that is,

$$\begin{aligned}
\sigma_c^2 &= NL(1 - 2^{-m})\frac{2^{-2b}}{12} \\
&= NL(1 - 2^{-m})\sigma_e^2
\end{aligned} \tag{3.21}$$

where $\sigma_e^2 = \frac{1}{12}2^{-2b}$ with products rounded to b bits, which are the finite register word length [108].

It is clear from above that mean square value of round-off noise of DWT coefficients is proportional to N, the number of points transformed and increases with increase in depth of decomposition.

A useful measure of accuracy of DWT coefficients is signal to noise ratio (SNR). The ratio of the signal power to the noise power is a useful measure of the relative strengths of the signal and the noise [108]. For rounding, the SNR of DWT coefficient at mth level is

$$\frac{\sigma_x^2}{\sigma_{dm}^2} = \frac{\sigma_x^2}{NL(1 - 2^{-m})\frac{2^{-2b}}{12}} = 12 \times 2^{2b} \frac{\sigma_x^2}{NL(1 - 2^{-m})} \qquad (3.22)$$

As we have seen that if the input signal exceeds the dynamic range of the quantization process, we must reduce the input amplitude to eliminate clipping. That is, we quantize samples $Ax(n)$ instead of $x(n)$, where $0 < A < 1$. Since the variance of $Ax(n)$ is $A^2\sigma_x^2$, the resultant SNR is

$$\begin{aligned}
SNR_{dm} &= 6.02\,b + 10.79 + 10\,\log_{10}(A^2\sigma_x^2) - 3.01\,v - 10\log_{10}(L) - 10\,\log_{10}(1 - 2^{-m}) \\
&= 6.02\,b + 10.79 + 10\log_{10}(\sigma_x^2) + 20\,\log_{10}(A) - 3.01\,v - 10\log_{10}(L) - 10\,\log_{10}(1 - 2^{-m})
\end{aligned}$$
$$(3.23)$$

As $0 < A < 1$, hence reducing the amplitude of the input to reduce clipping distortion reduces the SNR.

It is clear that SNR increases approximately 6.0 db for each bit added to the register length. In addition, it decreases by 3.01 db for increase of signal sample $N = 2^v$ to $N = 2^{v+1}$. The SNR decreases for increase in depth of decomposition level m. The SNR decreases with increase in filter tap length L.

The DWT algorithm, once transposed, can be used to implement an IDWT. It can be shown that DWT and IDWT require exactly the same number of operations (multiplications and additions) per point. Thus, total complexity for m level pyramid DWT algorithm is given by $4L(1 - 2^{-m})$ *mults/point*, and $4(L - 1)(1 - 2^{-m})$ *adds/point*. The variance of round-off errors in multiplication operations at the output of reconstructed signal is

$$\sigma_{ex'}^2 = 4NL(1 - 2^{-m})\frac{2^{-2b}}{12} \qquad (3.24)$$

and

$$\frac{\sigma_x^2}{\sigma_{ex'}^2} = \frac{\sigma_x^2}{4NL(1 - 2^{-m})\frac{2^{-2b}}{12}} = 3 \times 2^{2b} \frac{\sigma_x^2}{NL(1 - 2^{-m})} \qquad (3.25)$$

The signal to noise ratio is

$$
\begin{aligned}
SNR_{x'} &= 6.02\,b \; + \; 4.77 \; + 10\,\log_{10}(A^2\sigma_x^2) - 3.01\,v - 10\,\log_{10}(L) \; - \; 10\,\log_{10}(1 - 2^{-m}) \\
&= 6.02\,b \; + \; 4.77 \; + 10\,\log_{10}(\sigma_x^2) + 20\,\log_{10}(A) - 3.01\,v - \; 10\,\log_{10}(L) - 10\,\log_{10}(1 - 2^{-m})
\end{aligned}
$$

$$(3.26)$$

Chapter 4
PVM Implementation of DWT-Based Image Denoising

Abstract Recently users from both high-performance scientific community and general-purpose applications have shown keen interest in parallel processing due to its higher performance, lower cost, and sustained productivity [148]. To solve a computationally intensive problem efficiently on a cluster of existing computers, distributed computing involves a significantly lower cost factor [156]. Although it is difficult for a distributed computing user to achieve the computational capacity of large massively parallel processors (MPP), it is possible to solve large-size problems by combining a variety of distributed computing resources, connected by high-speed networks. This approach has advantages in terms of flexibility, scalability, and low cost. The advantage of using a cluster of workstations as the computational platform is that a cluster of a large number of workstations is easily available. A disadvantage is that there may be many users running unrelated tasks on the workstations so that the available computing resource for each task fluctuates in an unpredictable manner. Furthermore, communication between workstations is relatively slow. Although the performance of generalized cross-validation (GCV)–based threshold selection scheme is excellent, it is costly from CPU time viewpoint when implemented sequentially. In contrast to the traditional parallel approaches, which rely on specialized parallel machines, present work explores the potential of distributed systems for parallelism. The master/slave model is adopted for control of machines. This chapter is organized as follows. Section 4.1 presents background material and review of work in related area. Section 4.2 presents parallel algorithm of DWT. Basics of PVM programming has been discussed in Sect. 4.3.

Keywords PVM · Parallel algorithms · Speedup

K. K. Shukla and A. K. Tiwari, *Efficient Algorithms for Discrete Wavelet Transform*,   51
SpringerBriefs in Computer Science, DOI: 10.1007/978-1-4471-4941-5_4,
© K. K. Shukla 2013

4.1 Introduction

Parallelism is one of the oldest and most important techniques used to improve the performance of computing systems and has been applied extensively at virtually every level of system hierarchy.

A noise reduction (denoising) technique is basically method of removing the noise while retaining important features of the images. Various signal denoising schemes via *wavelet thresholding* or *shrinkage* have shown that wavelet thresholding for denoising has near optimal properties in the mean square error (MSE) sense and performs well in simulation studies of one-dimensional signal estimation [51, 52, 118, 147, 148]. Additive *Gaussian noise* (zero mean and standard deviation σ) filtering via thresholding wavelet coefficients was pioneered by Donoho [39–42, 148]. In this scheme, a wavelet coefficient is compared with a given threshold and is set to zero if its magnitude is less than the threshold; otherwise, is kept or modified (depending on the thresholding rule). The threshold acts as an oracle, which distinguishes between the insignificant coefficients possibly due to noise, and significant coefficients that encode important signal features. Signals with sparse or near sparse representation, where only a small subset of the coefficients represent all or most of the signal energy, are fit for thresholding. Wavelet transform is known to possess energy compaction properties, that is, a small number of coefficients at approximate level contain most of the signal energy. By thresholding, one essentially creates a region around zero where the coefficients are considered negligible. Outside this region, the thresholded coefficients are kept to full precision (that is without quantization).

The DWT applications are computationally intensive, and single-processor implementations are not suitable for current applications. They have been implemented in several parallel systems [52, 55]. In this chapter, computation of DWT coefficients on a network of workstation clusters is explored. There are several standards for distributed computing including PVM, P4, Express, MPI, and Linda. A comprehensive summary of the recent developments in distributed system design is presented in [118], where PVM is viewed as the existing de facto standard for distributed computing and MPI is being considered as the future message-passing standard. In PVM environment, a programmer can exploit distributed computing across a wide variety of computer types. Thus, PVM constitutes a heterogeneous network of computers to appear as one single concurrent computational engine—a large virtual machine as indicated by the name. The PVM system has got acceptability in high-performance scientific computing community due to its simple concept and simple but complete programming interface [55]. It facilitates simple program structures to be implemented in an intuitive manner. Each application consists of a collection of cooperating tasks, which access PVM resources through a library of standard interface routines. These routines allow the initiation and termination of the tasks across the network as well as communication and synchronization between tasks.

4.2 Multicomputer Network

Multicomputer systems are nonshared memory architectures where processors communicate through message passing. These systems are characterized by asynchronous parallelism, and synchronization is data driven. It is easier for one to solve a particular class of problem efficiently on a nonshared memory machine based on specific or fixed interconnection network by exploiting topologies of their choice. The message-passing multicomputers are lattices of processing elements (PE) nodes connected by a message-passing network. The basic computational paradigm is that of concurrency of processes, where processes are instances of programs.

4.2.1 Parallel Algorithm

The DWT may be computed recursively as a series of convolutions and decimations [52]. At each scale level j, an input sequence $S^{j-1}(n)$ is fed into a *low-pass* and *high-pass* filters $G(z)$ and $H(z)$, respectively. The output from the *high-pass* filter $H(z)$ represents the detail information in the original signal at the given scale j, denoted by $W^j(n)$. The output from the *low-pass* filter $G(z)$ represents the remaining information in the original signal and is denoted by $S^j(n)$. $G(z)$ and $H(z)$ are the z-transform of the set of coefficients representing the scaling function $\varphi(t)$ and the wavelet function $\psi(t)$, respectively. The sequential DWT algorithm may be expressed as

$$S^j(n) \quad = \quad \sum_k S^{j-1}(k)\, G\,(2n-k) \tag{4.1}$$

$$W^j(n) \quad = \quad \sum_k S^{j-1}(k)\, H\,(2n-k) \tag{4.2}$$

which can be further simplified as

$$S^j = S^{j-1}G \text{ and } W^j = S^{j-1}H \tag{4.3}$$

The equation for reconstruction is given by

$$S^{j-1} = S^j G^* + W^j H^* \tag{4.4}$$

where H^* and G^* are adjoint of the operators H and G. The algorithm follows a half pyramid structure. The parallel algorithms are developed as processes that can be mapped on to a certain number of processors with a fixed topology.

4.2.1.1 One-Dimensional Decomposition algorithms

Following are the nested loop one-dimensional DWT algorithm (sequential version):

1. $S^0(n) = X(n);\ S^j(n) = 0\ \forall j;\ \forall n$
2. for $j = 1$ to J
3.   for $n = 1$ to $M/2^j$
4.   for $k = \max(1;\ 2n-L + 1)$ to $2n$
5. $S^j(n) = S^j(n) + S^{j-1}(k)G(2n-k)$
6. $W^j(n) = W^j(n) + S^{j-1}(k)H(2n-k)$
7.   endfor(k)
8.   endfor(n)
9. endfor(j)
10. output $= S^J(n),\ W^j(n),\ j = 1, 2, \ldots, J.$

The parallel version of above algorithm may be derived with assumption of decomposition into a coarser approximation and finer detail at each scale as a process. Let, j be the current scale and length of the signal is a power of two.

Process 0 at root
For number of [data packets]
Initialize data packets to x_o's

1. Send $[x_o$'s$]$ to process 1
   processes $j = 1$ to *maxscale*
2. Load H and G filters for each process
3. Receive $[S^{j-1}]$
4. Compute coefficients (S^j's and W^j's)
5. Send S^j's to process $j + 1$ (while $j <$ maxscale)

The coefficients at each stage are returned to the root processor, taking the shortest path.

Similarly, we can derive algorithm for one-dimensional reconstruction (i.e., inverse DWT) as follows:

Process $j = maxscal$

1. Send S^j's and W^j's
   processes $j = $ *maxscale to* 1 *step*-1
2. Receive S^j's and W^j's
3. Load H^* and G^* filters
4. Compute $[S^{j-1}]$
   Send S^{j-1} to process $j-1$ (while $j > 1$)

4.2.2 *Timing Consideration*

The DWT computation involves recursive convolution and decimation. As is clear from fore-cited sequential and parallel algorithms, coefficient computation of next level is possible only after coefficients availability from previous level. Thus, the timing consideration of DWT decomposition in a multicomputers system is little involved one. Let us assume that the serial execution time for the sequential decomposition algorithm on single-computer system is given by

$$T_s = Tn \tag{4.5}$$

where T is the time to perform one computation (an addition or multiplication) and n is the total number of computation operations.

To obtain the parallel execution time, the total number of computations involved in decomposition may be divided into two parts, that is,

$$n = n_c + n_{nc} \tag{4.6}$$

where n_c is the number of computation in parallel execution that needs communication from other computers and n_{nc} are the number of computations that need no communication. Thus, the parallel execution time T_p is given by

$$T_P = T_x + T_y + T_s/P \tag{4.7}$$

where

$$T_x = T_{cb} \, n_c \, L \tag{4.8}$$

and

$$T_y = T_{cs} \, B \, n_c \, L \tag{4.9}$$

where T_{cb} is the time to begin a communication, T_{cs} is the time taken to send a data B (in bytes) over a link, and L is the path or number of links through which communication occurs. Thus, the total communication time is given by

$$T_c = T_x + T_y \tag{4.10}$$

The parallel time is the sum of the time for communications and the reduction in serial time using P processors,

$$T_P = T_c + T_s/P \tag{4.11}$$

4.3 Speedup Using PVM

Our algorithm belongs to the *master–slave* type. The computational platform selected here is a cluster of loosely connected workstations (the machines). The individual slave algorithm will be executed on a separate workstation. The master algorithm does not require many computing resources to execute and may share the computing resources of a machine already executing a slave algorithm. Direct communications are allowed only between master and slave algorithms [148]. Direct communications between slave algorithms are not allowed. Due to the fact that in loosely connected environment, the communication speed between machines is slow; hence, the communications between master and slave are kept to a minimum. The master performs initialization and assigning jobs to slaves. Each of the slaves performs DWT decomposition, threshold selection, and image denoising as per defined rule, inverse DWT, and after completing the task reports to the master.

The PVM computing model is shown in Fig. 4.1a. An architectural view of the PVM system, highlighting the heterogeneity of the computing platforms supported by PVM, is shown in Fig. 4.1b [55]. The PVM software provides a flexible parallel computing environment. It supports concurrent execution on loosely coupled network of PE in addition to message-passing model of synchronization. In PVM framework, a parallel program can be developed in an efficient way without any specific hardware requirements.

The advantage of parallel implementation using PVM on a cluster of workstations lies in effective utilization of available workstations. The image to be denoised is divided into subimages and distributed over nodes on the network. In the present investigation, authors have adopted master/slave (or host–node) parallel programming paradigm [55]. The software developed includes a master program and slave programs. The master program controls the data processed by a number of slave programs. It can run on one machine and spawn copies of the slave programs to any number of nodes in the network. The master program is responsible for process spawning, initialization, collection, and display of results and timing of functions. The slave program performs the actual computation involved.

The followings are in brief description of the master and slave control programs:

Master program:

1. Initialize the system
2. Locate a memory array for the test data
3. Read the test image into the memory array
4. Partition the test data into number of subtest data
5. Obtain task identifier (TID) by registering in PVM
6. Spawn copies of slave programs on the slave machines
7. Send control messages to slave machines
8. Send subtest data to slave machines
9. Wait for the slave machine to report back
10. Collect data from the slave machines for final output.

Fig. 4.1 PVM system overview [55]. **a** PVM computation model. **b** PVM architectural overview

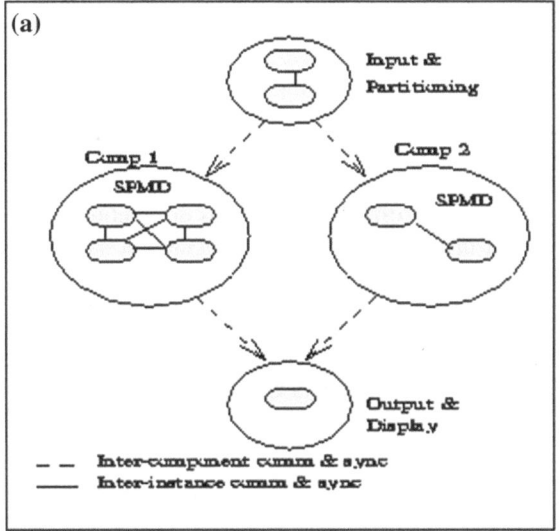

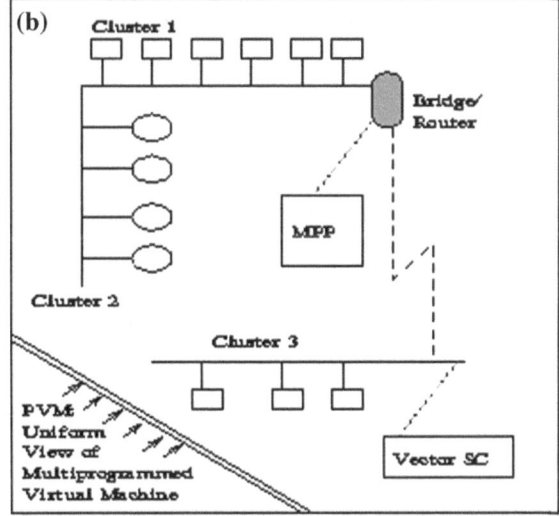

Slave program:

1. Initialize the data structure for the processed test data segments
2. Obtain its TID by registering in PVM process
3. Wait for data to be sent from the master
4. Receive the data and perform the specified operation
5. Inform the master machine when the result data are ready
6. Send the data to the master
7. Exit from the PVM process.

4.3.1 Workload Allocation

The PVM computing model supports both data decomposition and function decomposition. As most of the problems involve computational operation or transformations on one or more data structure, thus it is possible that these data structures may be divided and operated upon for problem solving. Dividing the job on the basis of different operations or function is called function decomposition. In present investigation for experimentation data, decomposition is performed. The main steps are outlined as follows:

1. Read the test image, filter coefficients, and level of decompositions into master node.
2. Segment the test data in master node and send each subimage to individual slave node along with filter coefficients and level of decomposition required.
3. Perform DWT decomposition at each slave under coordination of the master node.
4. Compute threshold value.
5. Process the coefficients with *soft* thresholding rule and perform inverse DWT.
6. Collection of results by master node from slaves.

Next, the performance measures used in our experiment speedup and efficiency are defined.

4.3.2 Speedup Factor

One of the most important criteria and a very commonly used factor, determining the usefulness of a parallel algorithm, is the speedup factor. The *speedup factor S* is defined as the ratio of wall clock execution time needed whether only one processor is used to that needed when P processors are used in parallel [148].

The speedup achieved by using a parallel network over a single processor is given by

$$S = \frac{T_s}{T_P} = \frac{T_s}{T_c + \frac{T_s}{P}} = P\left(\frac{1}{P\frac{T_c}{T_s} + 1}\right) \tag{4.12}$$

where T_c is the inter-processor communication time.

It is evident from above that the product of the communication time and number of processors should be minimal compared to serial execution time to obtain optimum speedup. The serial execution time is a function of the number of operations, which in turn increases with increase in complexity of algorithm. A high speedup factor indicates effective use of processors. For most parallel algorithms, the *speedup factor* is less than P when solving a problem using P processors.

4.3.3 *Efficiency*

The efficiency is an important measure for the performance of a parallel algorithm. It indicates the effectiveness of the processors [148]. The efficiency E is given by

$$E = \frac{S}{P} \tag{4.13}$$

As the efficiency is a ratio of speedup factor and number of processors, it is also called the *normalized speedup factor*.

Chapter 5
DWT-Based Power Quality Classification

Abstract This chapter presents application of DWT and fuzzy set theory to classification of power quality problems. The system uses discrete wavelet transform as linear filters for preprocessing and fuzzy expert system for feature extraction and classification. The signal under test (electrical current or voltage for power quality study) is processed through a DWT decomposition block to generate the feature extraction curve. Then, a fuzzy logic–based inference engine utilizing these features as inputs is implemented for decision making. The DWT level and energy information from the feature extraction curve are passed through a diagnostic module that computes the truth value of the signal combination and determines the class to which the signal belongs. Also presented are comparative performances of fuzzy inference engine with various defuzzification procedures. The proposed scheme has been validated for both *Mamdani-type* and *Sugeno-type* fuzzy inference engines. The proposed scheme is much simpler and powerful than currently available power quality (PQ) classification schemes. The organization of the chapter is as follows. Section 5.1 presents background material of the subject. Section 5.2 presents a general introduction to application of DWT in PQ classification. Application of fuzzy inferencing and DWT for monitoring PQ issues has been dealt with in Sect. 5.3, which covers in detail the results related to PQ classification with DWT system. Finally, Sect. 5.4 presents conclusions.

Keywords Power quality · Fuzzy inference · Feature detection · Decision support system

5.1 Introduction

Many techniques have been used to monitor power quality (PQ) problems. Disturbance analyzers have been developed to measure the problem in electrical signals [5]. The fast Fourier transform (FFT) calculation capability has been added

K. K. Shukla and A. K. Tiwari, *Efficient Algorithms for Discrete Wavelet Transform*,   61
SpringerBriefs in Computer Science, DOI: 10.1007/978-1-4471-4941-5_5,
© K. K. Shukla 2013

to some disturbance analyzers to get a clear picture of the harmonic content within the distorted signal [44]. The commonly used method for detecting PQ disturbances is based on a point-to-point comparison of adjacent cycles [44]. The drawback of this approach is that it fails to detect disturbances that appear periodically, such as flattop and phase controlled load wave shape disturbances. Another approach to detect and identify disturbances is based on neural networks [73]. Ghosh et al. [54] have studied both multilayered feed-forward and time delay ANN architectures for power system disturbance waveform classification. The neural network architectures suffer from the requirement of large number of training cycles and resultant computational burden in sequential implementation.

Wavelet analysis is proposed as a new tool for monitoring PQ problems. All the recently proposed approaches [65, 112, 119] utilize the localization property of the wavelet transform. However, there is no real classification of different types of PQ problems and quantification of the magnitude of these disturbances. Dash et al. [32] proposed a hybrid scheme using a *Fourier linear combiner* and a *fuzzy expert system* for the classification of transient disturbance waveforms in a power system. However, Fourier transform is unable to localize events in both frequency and time domains.

All the above-mentioned approaches dealt with PQ problems. However, none of them presents a methodology that can be used to classify or measure different PQ problems. In this chapter, application of discrete wavelet transform, as a powerful tool for detecting, localizing and for classification and quantification of power signal disturbances, has been investigated. In the current investigation, DWT has been used as preprocessor and afterward processing of wavelet coefficients by fuzzy logic has been carried out. Wavelet preprocessing permits reduction of the dimensionality of the problem.

5.2 DWT in Feature Detection and Extraction

In this section, we summarize the application of the wavelet transform in detecting and extracting PQ disturbances. In the present investigation, we utilize a dyadic-orthonormal wavelet transform with Daubechies' wavelet with four-coefficient filter [107]. Santoso et al. [120, 121] proposed a 12-scale DWT decomposition of electrical signals to ensure that all disturbance features in both high and low frequency are extracted. First, scale signal has frequency range of $f/2$–$f/4$, where f is the sampling frequency of the time domain disturbance signal. The second-, third-, fourth-, fifth-, and higher-level signals have frequency ranges of $f/4$–$f/8$, $f/8$–$f/16$, $f/16$–$f/32$, $f/32$–$f/64$, respectively. Figure 5.1 plots two-level DWT decomposition and reconstruction system. The system can easily be extended for J levels. The wavelet transform is useful in detecting and extracting disturbance features of various types of electric PQ disturbances because it is sensitive to signal irregularities but insensitive to the regular-like signal behavior.

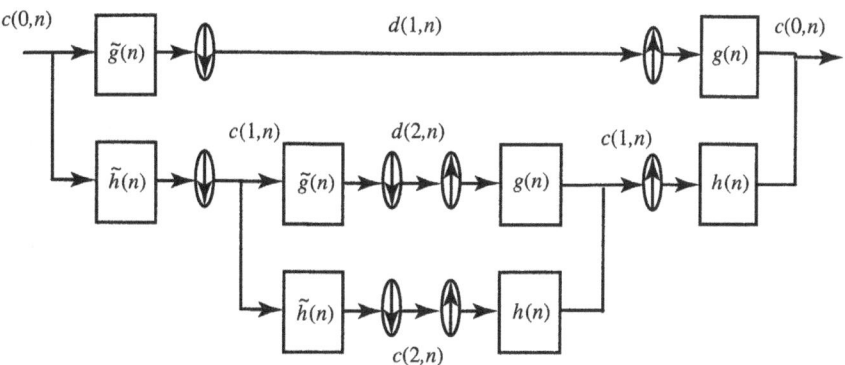

Fig. 5.1 DWT (*two-level*) decomposition and reconstruction system

5.2.1 Detection and Localization of PQ Disturbances

In the present investigation, disturbance detection and localization are performed in the wavelet domain rather than the time or frequency domain. Classification of PQ problems in the present investigation is based upon Parseval's theorem, which states that if the used wavelets form an orthonormal basis and satisfy the admissibility condition [113], then the energy of the distorted signal is equal to the energy in each of the expansion coefficients, that is,

$$\int |f(t)|^2 = \sum_{k=-\infty}^{\infty} |c(k)|^2 + \sum_{j=0}^{\infty} \sum_{k=-\infty}^{\infty} |d_j(k)|^2 \tag{5.1}$$

The energy is partitioned in time by k and in scale by j in the wavelet domain. Thus, the energy of the signal will be partitioned differently for different PQ problem.

5.2.1.1 Detection, Localization, and Classification

Using the localization property gained from finer resolution levels, a time–frequency plot of the distorted signal is generated. In addition, a plot of energy contained in DWT coefficients at different levels is computed, which represents the energy distribution of the distortion at different frequency bands. Using this information, one can classify different PQ problems. Thus, feature detection and extraction proceed as follows:

1. Using DWT, decompose the signal under test (Number of levels is selected to cover the highest frequency band of interest).
2. Find energy of signal at different resolution levels.

3. Construct feature extraction curve by plotting energy at each level.
4. First, detail version present time information of transient event.

The proposed algorithm is tested on various PQ problems, for example, sag, swell, interruption, and harmonic distortion (both higher order and subharmonic). The results are generated with Matlab [93]. Mother wavelet used is Daubechies Db4 [37].

Figures 5.2, 5.3, 5.4, 5.5, 5.6, 5.7, and 5.8 illustrate efficiency of the proposed algorithm in detecting and classifying various PQ problems, viz. pure sine wave, sag in sine wave, swell in sine wave, interruption in sine wave, outage in sine wave and harmonics. Figures 5.2, 5.3, 5.4, 5.5, 5.6, 5.7, and 5.8 plots

1. Signal under test,
2. Three finer decomposition levels, and
3. Feature extraction curve (plot between decomposition level on "x"-axis and energy at different resolution on "y"-axis).

Figure 5.9 plots a generalized feature extraction curve derived based on experimentations. The resultant classification rules are so simple for the operator to detect, localize, and classify different PQ problems based on the proposed feature extraction curve. The variation in three different zones of the feature extraction curve, peak (voltage sag and swell), lower left part (higher harmonic), and lower right (subharmonic), provides sufficient information for classifying different PQ problems. However, looking onto feature extraction curve many a time, it will be difficult for operator to distinguish between different PQ problems impended in the signal simultaneously and to predict about the magnitude of particular problems. To overcome the limitations, *fuzzy expert system* [43]–based PQ classification is proposed in next section.

5.2.2 Expert System in PQ Classification

Expert systems have been defined as "an intelligent computer program that uses knowledge and inference procedures to solve problems that are difficult enough to require significant human expertise for their solution" [90]. In a procedural (conventional) program, the user must specify exactly how an algorithmic pre-defined number of steps, the solution to the problem is to be achieved. In expert system (ES), the user specifies the goal and gives the system the ability to reason (infer), and the system decides how to accomplish the goal.

5.2.2.1 Fuzzy Logic

Crisp variables and crisp knowledge are elements in the knowledge domain that have an exact *truth table*, either True or False. Fuzzy logic and fuzzy sets are tools

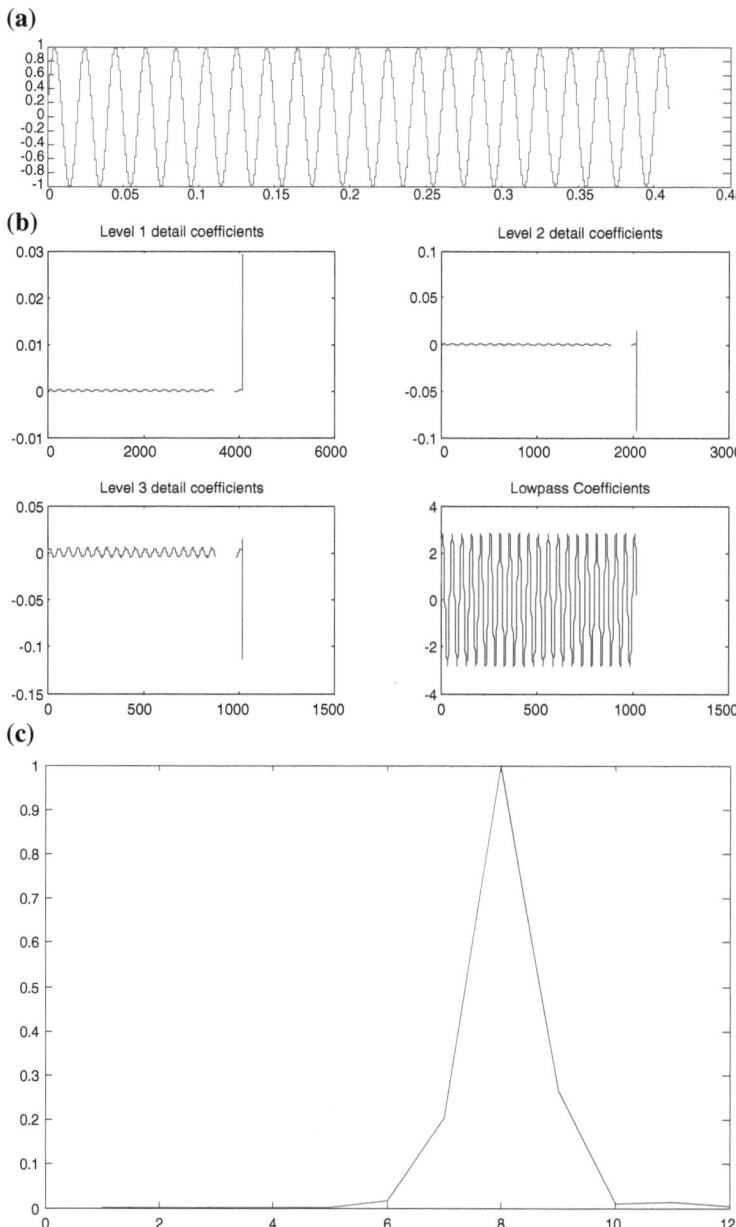

Fig. 5.2 Pure sine wave, DWT decomposition, and feature extraction curve. **a** Pure sinusoidal waveform (x-axis time, y-axis magnitude). **b** DWT decomposition of signal under test (x-axis coefficient number, y-axis magnitude). **c** Derived feature extraction curve (x-axis level, y-axis energy)

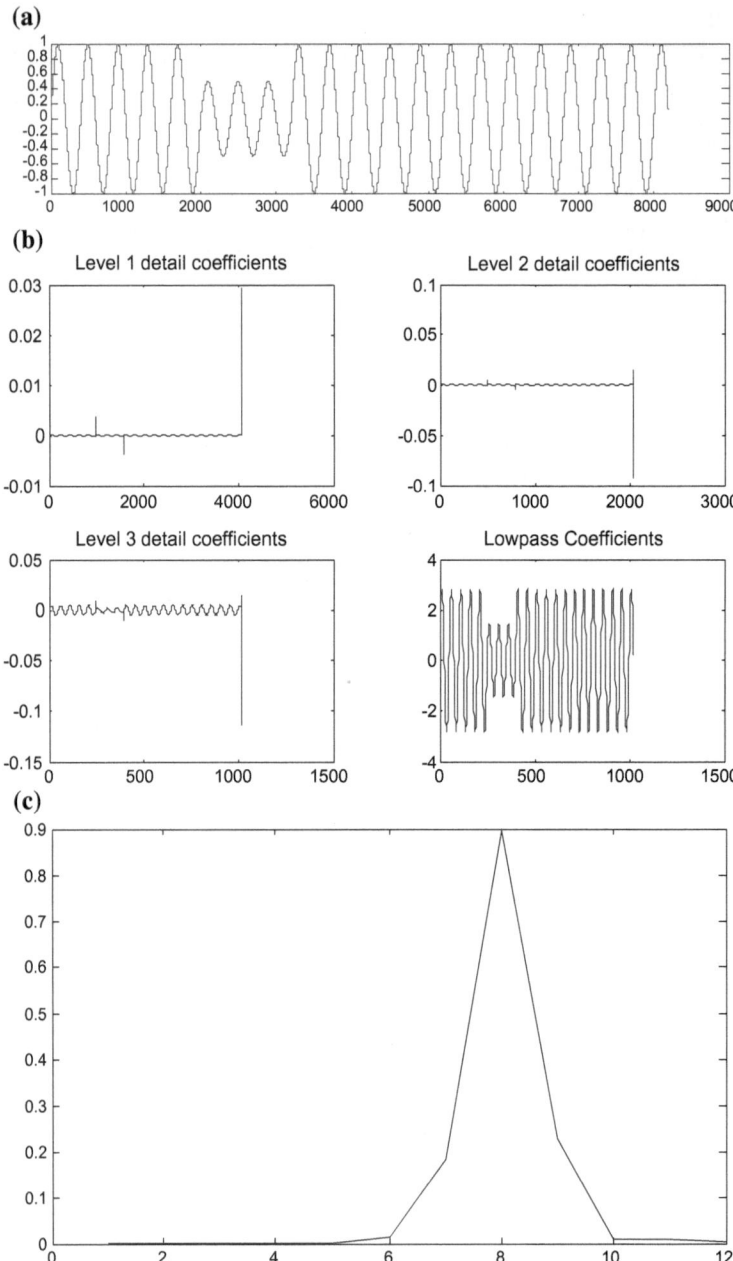

Fig. 5.3 Sag in sine wave, DWT decomposition, and feature extraction curve. **a** Test signal with Sag in *sine wave* (x-axis time, y-axis magnitude). **b** DWT decomposition of signal under test (x-axis coefficient number, y-axis magnitude). **c** Derived feature extraction curve (x-axis level, y-axis energy)

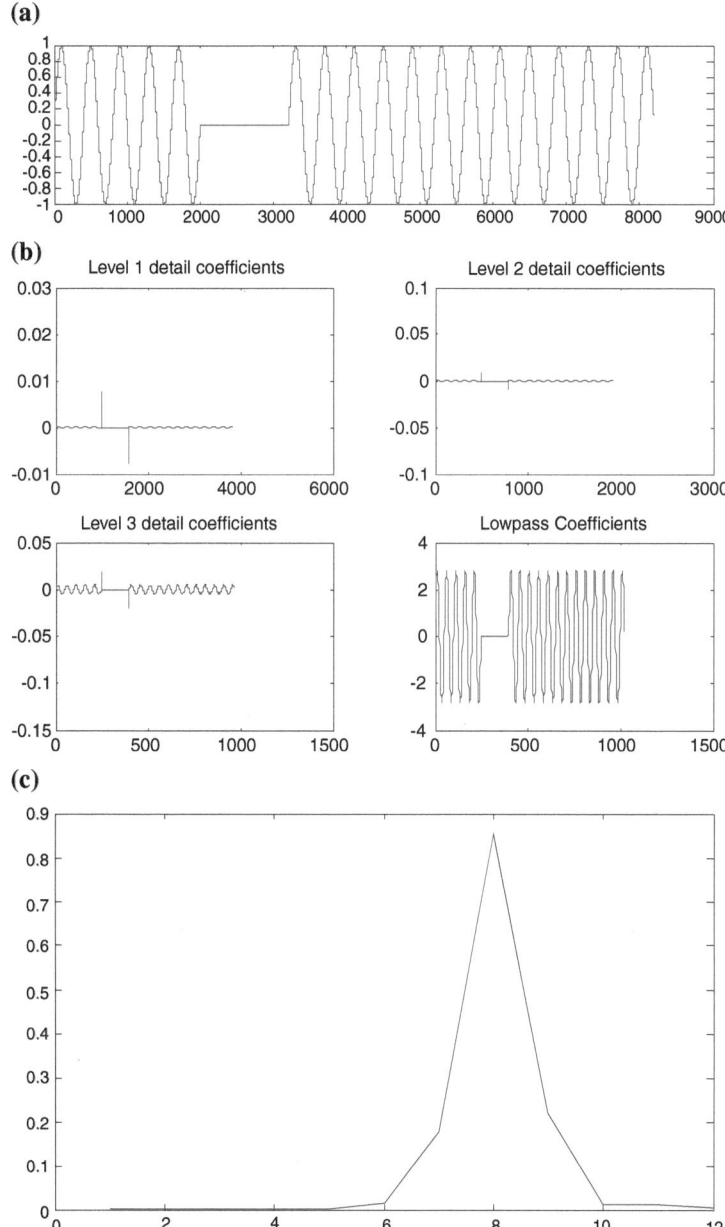

Fig. 5.4 Outage sine wave, DWT decomposition, and feature extraction curve. **a** Test signal with outage in *sine wave* (x-axis time, y-axis magnitude). **b** DWT decomposition of signal under test (x-axis coefficient number, y-axis magnitude). **c** Derived feature extraction curve (x-axis level, y-axis energy)

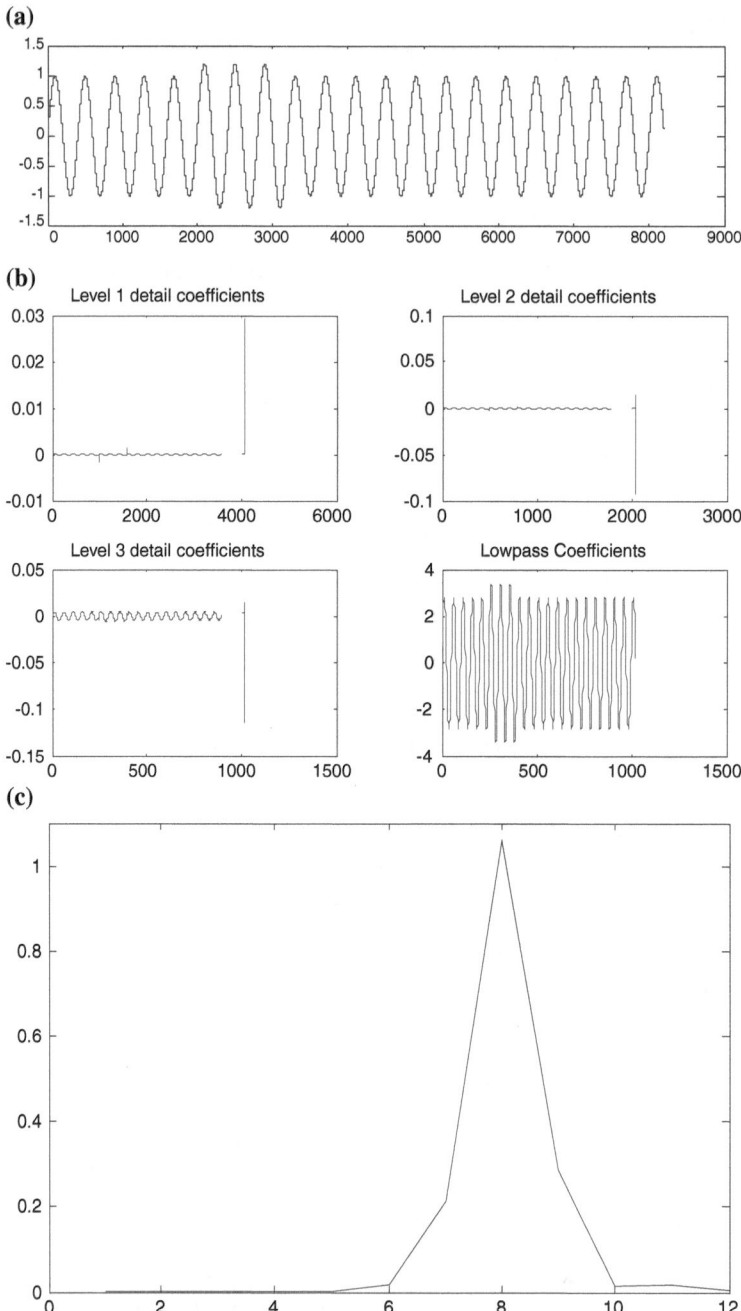

Fig. 5.5 Swell in sine wave, DWT decomposition, and feature extraction curve. **a** Test signal with swell in *sine wave* (x-axis time, y-axis magnitude). **b** DWT decomposition of signal under test (x-axis coefficient number, y-axis magnitude). **c** Derived feature extraction curve (x-axis level, y-axis energy)

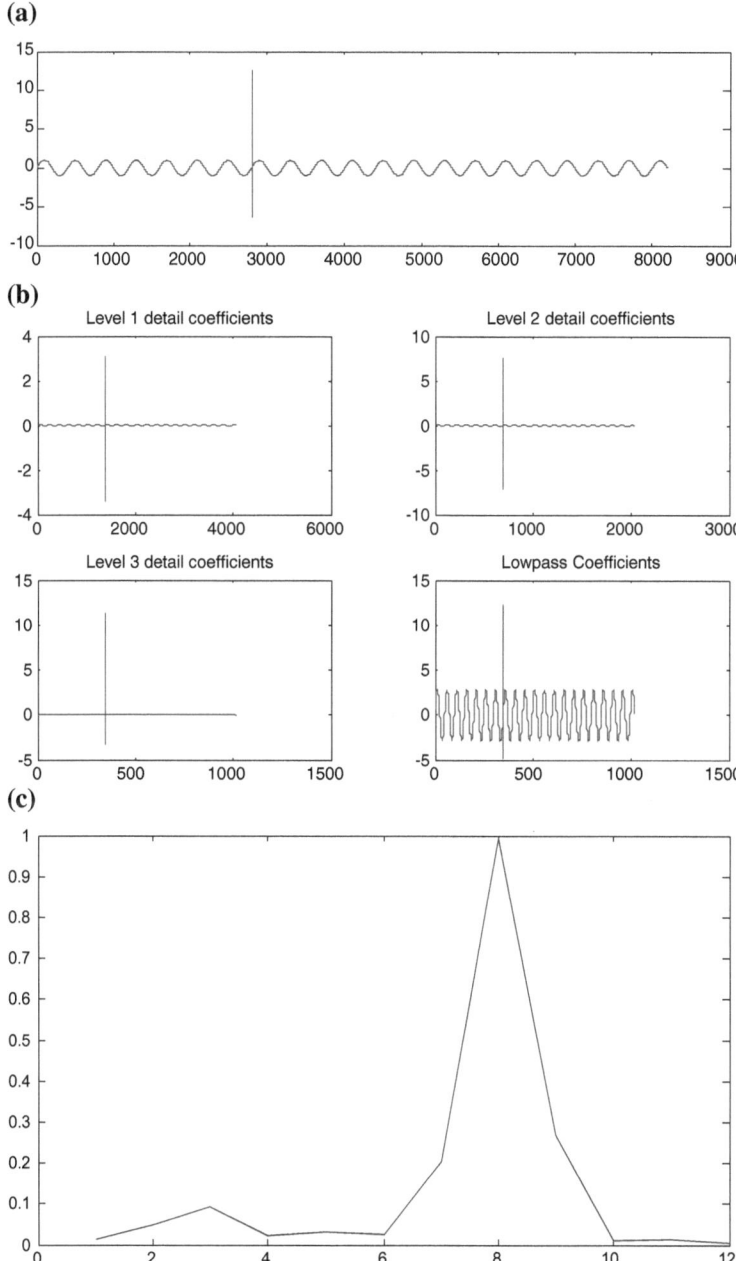

Fig. 5.6 Surge in sine wave, DWT decomposition, and feature extraction curve. **a** Test signal with surge in *sine wave* (x-axis time, y-axis magnitude). **b** DWT decomposition of signal under test (x-axis coefficient number, y-axis magnitude). **c** Derived feature extraction curve (x-axis level, y-axis energy)

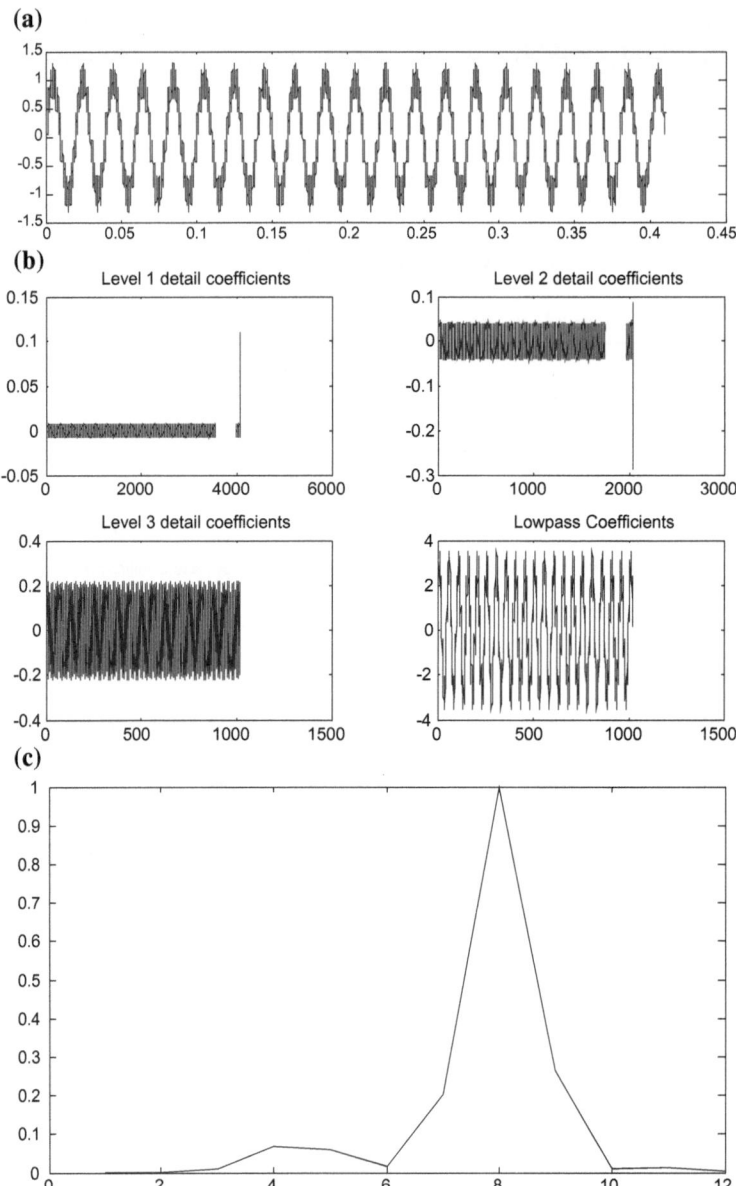

Fig. 5.7 Higher harmonics in sine wave, DWT decomposition, and feature extraction curve. **a** Test signal with higher harmonics in *sine wave* (x-axis time, y-axis magnitude). **b** DWT decomposition of signal under test (x-axis coefficient number, y-axis magnitude). **c** Derived feature extraction curve (x-axis level, y-axis energy)

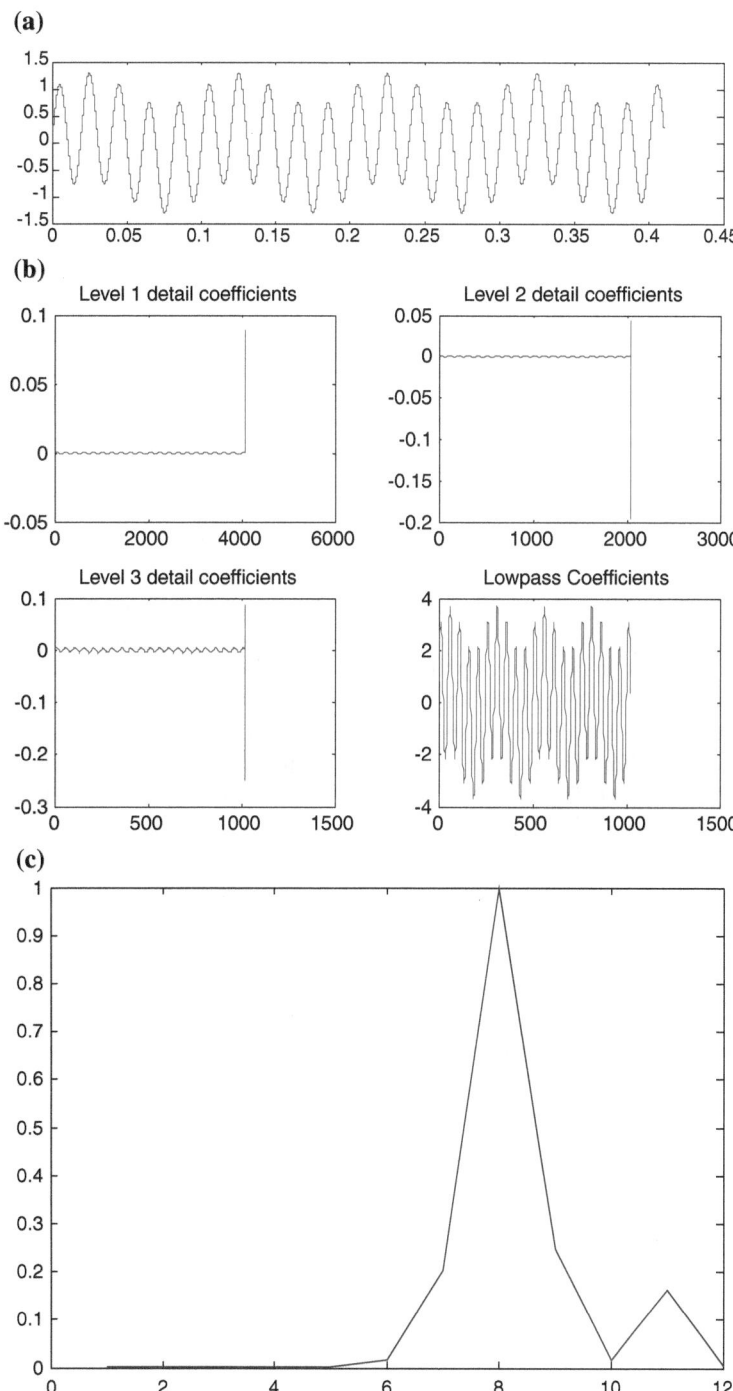

Fig. 5.8 Subharmonics in sine wave, DWT decomposition, and feature extraction curve. **a** Test signal with subharmonic in *sine wave* (x-axis time, y-axis magnitude). **b** DWT decomposition of signal under test (x-axis coefficient number, y-axis magnitude). **c** Derived feature extraction curve (x-axis level, y-axis energy)

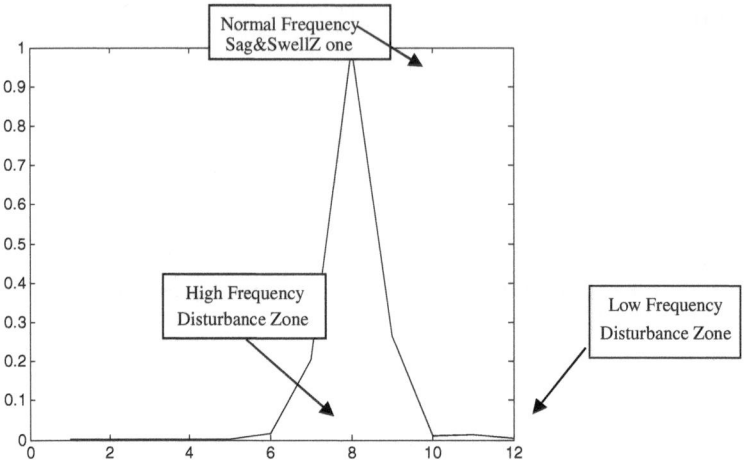

Fig. 5.9 Generalized PQ feature extraction curve (x-axis level, y-axis energy)

for expressing and operating on knowledge that is imprecise, or where the interpretation is highly subjective and depends strongly on context or human opinion. The fuzziness of a property lies in the lack of well-defined boundaries of the set of objects to which this property applies. In fuzzy set theory, "normal" sets are called crisp sets, in order to distinguish them from fuzzy sets. For any crisp set C, it is possible to define a characteristics function $\mu_c = U \rightarrow \{0, 1\}$. In fuzzy set, the characteristics function is generalized to a membership function that assigns to every $u \in U$ a value from the unit interval $\{0, 1\}$ instead from the two elements set $\{0, 1\}$ [43]. The set that is defined on the basis of such an extended membership function is called a fuzzy set [130].

Figure 5.10 shows the basic configuration of a fuzzy logic classifier (FLC), which comprises four principal components: a fuzzification interface, a knowledge base (KB), decision-making logic, and a defuzzification interface.

A defuzzification strategy is aimed at producing a nonfuzzy action that best represents the possibility distribution of an inferred fuzzy data. Unfortunately, there is no systematic procedure for choosing a defuzzification strategy. Followings are the commonly used defuzzification strategies for which proposed classification methodology has been tested [81, 88] Fig. 5.11:

1. Centroid of area method (CoA).
2. Bisector of area method.
3. Smallest of maximum method (SoM).
4. Largest of maximum method (LoM).
5. Mean of maximum method (MoM).

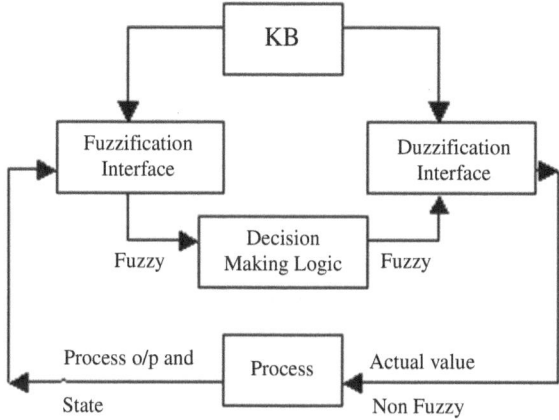

Fig. 5.10 Basic configuration of a fuzzy logic classifier

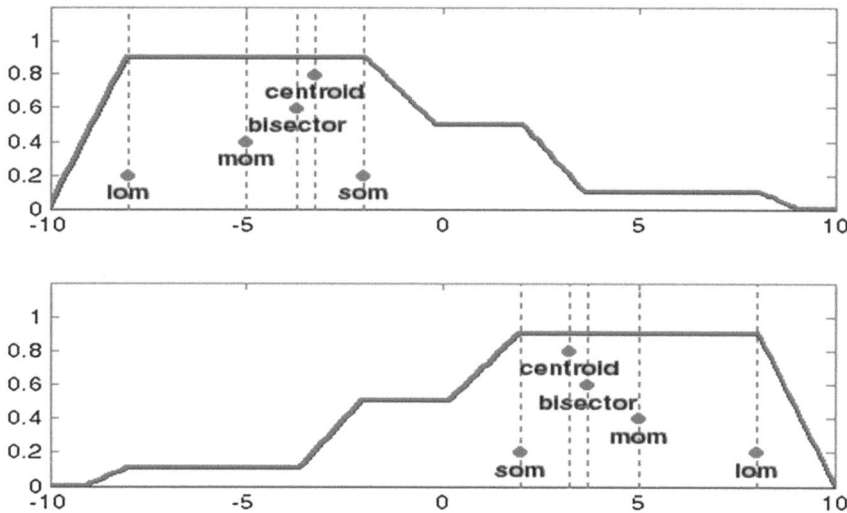

Fig. 5.11 Diagrammatic representation of various defuzzification strategies

An ideal defuzzification method should satisfy following criteria:

1. Continuity: A small change in input should not result in large change in output;
2. Intuition: A crisp value should represent the fuzzy set from an intuitive viewpoint (e.g., maximum membership grade);
3. Disambiguity;
4. Plausibility;
5. Computational Complexity; simple (to enable real-time hardware's).

5.2.2.2 Choice of Membership Functions

The linguistic values taken by variables in the rule antecedent, rule consequent, and the symbolic representation of the rules are good enough to allow some qualitative analysis concerning the stability of fuzzy logic–based classification system. However, for the needs of a quantitative description of behavior of the system, involving the quantitative computation of the output, one needs a quantitative interpretation of the meaning of the linguistic values. For computational efficiency, efficient use of memory, and performance analysis needs, a uniform representation of the membership functions is required. This uniform representation can be achieved by employing membership functions with uniform shape and parametric, functional definition. In present investigation, we had selected membership function [43] of "generalized bell-shaped" defined by

$$f(x; a, b, c) = \frac{1}{1 + \left| \frac{(x-c)}{a} \right|^{2b}} \tag{5.2}$$

where the parameter b is usually positive. The parameter c locates the center of the curve, and parameter a locates flat top portion of the curve. Figure 5.12 shows a plot of generalized bell shaped with specified parameters a, b, and c.

5.2.2.3 Derivation of Fuzzy Rules

There are four modes of derivation of fuzzy rules, as reported in [43]. These four modes are not mutually exclusive, and it seems likely that a combination of them would be necessary to construct an effective method for the derivation of fuzzy control rules.

1. Expert experience and control engineering knowledge,
2. Based on operator's control action,
3. Based on the fuzzy model of the process, and
4. Based on learning.

   In the present investigation, we had derived fuzzy rules based on mainly (1) and (2).

5.2.2.4 Fuzzy Inference Systems

Fuzzy inference is the process of formulating the mapping from a given input to an output using fuzzy logic. The mapping then provides a basis from which decisions can be made, or patterns discerned. The process of fuzzy inference involves all of the pieces that are described in the previous sections: membership functions, fuzzy logic operators, and if-then rules. There are two types of fuzzy inference systems that can be implemented: *Mamdani-type* Fig. 5.13 and *Sugeno-type* Fig. 5.14. These two types of inference systems vary somewhat in the way outputs are determined [43, 88, 130].

Here are some final considerations about the two methods.

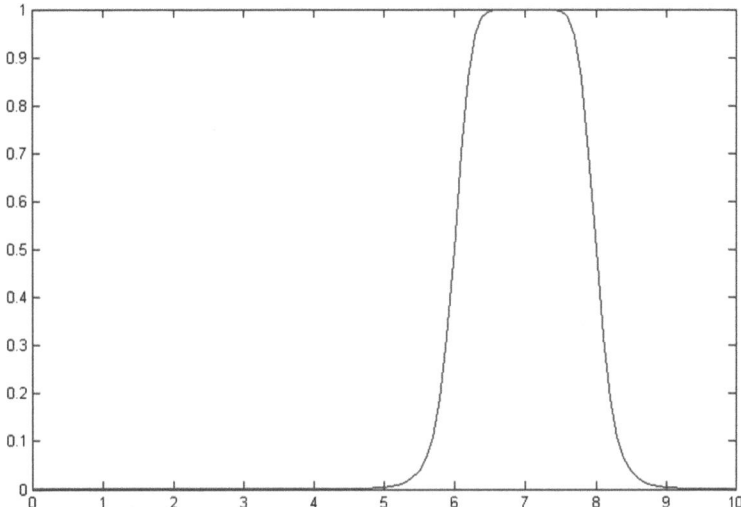

Fig. 5.12 Plot of generalized *bell-shaped* MF with $a = 1$, $b = 4$, $c = 7$

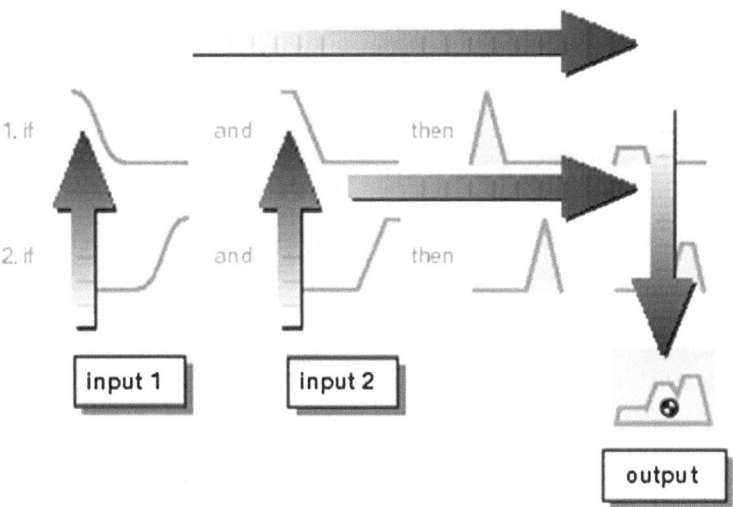

Fig. 5.13 Interpreting the fuzzy inference diagram (Mamdani's method)

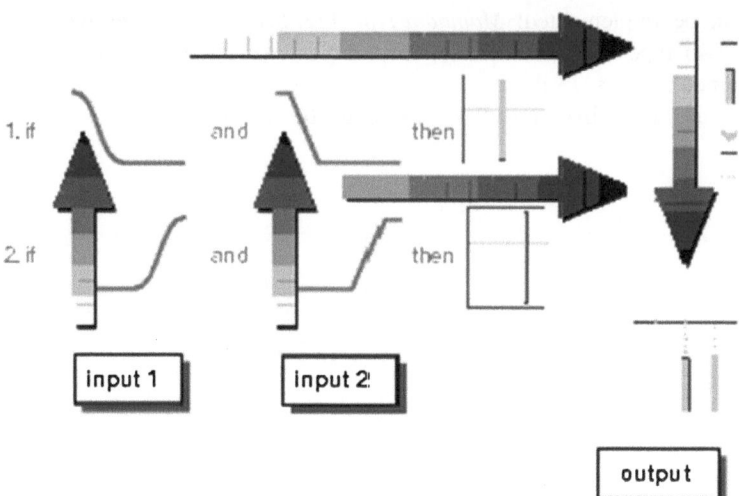

Fig. 5.14 Interpreting the fuzzy inference diagram (Sugeno's method)

Advantages of the Mamdani Method

- It is intuitive.
- It has widespread acceptance.
- It is well suited to human input.

Advantages of the Sugeno Method

- It is computationally efficient.
- It works well with linear techniques.
- It works well with optimization and adaptive techniques.
- It has guaranteed continuity of the output surface.
- It is well suited to mathematical analysis.

5.3 Results and Discussion

5.3.1 Application of Fuzzy Expert System in PQ Classification

For classifying the PQ problems, five fuzzy sets are chosen from the DWT levels (lv) designated as surge (fast transient/surges), higher order harmonic (hh), fundamental waveform (fn), subharmonic component (sb), and dc-offset (dc) (Fig. 5.15). In a similar way, five fuzzy sets are chosen for the amplitude of the feature extraction curve (en), designated as interruption/outage (intrp), lower peak (lv), peak corresponding to fundamental (nm), higher peak (hv), and surge (sg) (Fig. 5.16). The fuzzy set corresponding to particular PQ is as plotted in Fig. 5.17.

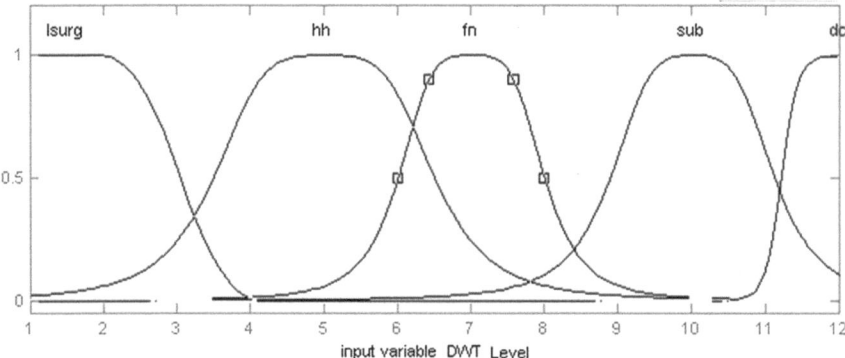

Fig. 5.15 Membership function for DWT levels of feature extraction curve

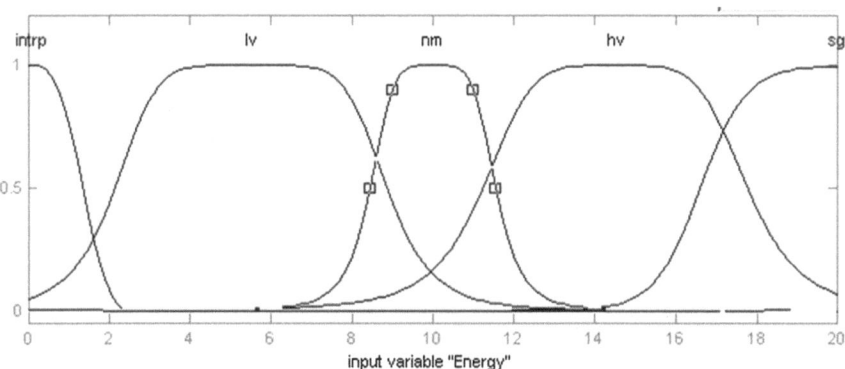

Fig. 5.16 Membership function for energy in feature extraction curve

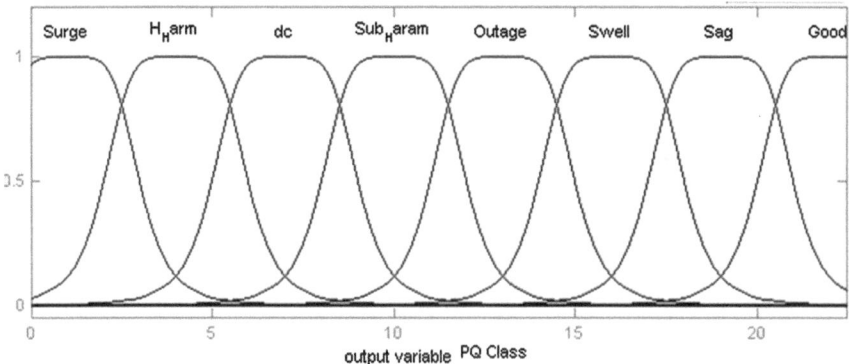

Fig. 5.17 Membership function for PQ classes

Table 5.1 Rule base for the fuzzy decision support system

Rule 1	If (DWT_level is fn) and (energy is nm)	then (PQ_class is good)
Rule 2	If (DWT_level is fn) and (energy is lv)	then (PQ_class is sag)
Rule 3	If (DWT_level is fn) and (energy is hv)	then (PQ_class is swell)
Rule 4	If (DWT_level is fn) and (energy is intrp)	then (PQ_class is outage)
Rule 5	If (DWT_level is sub) and (energy is lv)	then (PQ_class is subharmonic)
Rule 6	If (DWT_level is dc) and (energy is lv)	then (PQ_class is dc)
Rule 7	If (DWT_level is hh) and (energy is lv)	then (PQ_class is H_harmonic)
Rule 8	If (DWT_level is isurge) and (energy is nm)	then (PQ_class is surge)
Rule 10	If (DWT_level is isurge) and (energy is hv)	then (PQ_class is surge)
Rule 11	If (DWT_level is isurge) and (energy is sg)	then (PQ_class is surge)

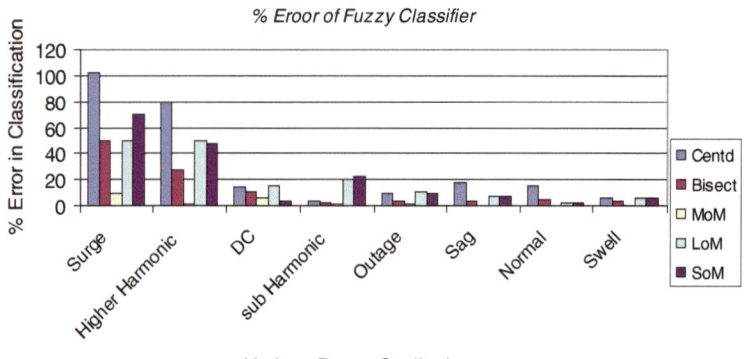

Fig. 5.18 Classification error with different defuzzification strategy (Mamdani's fuzzy inference)

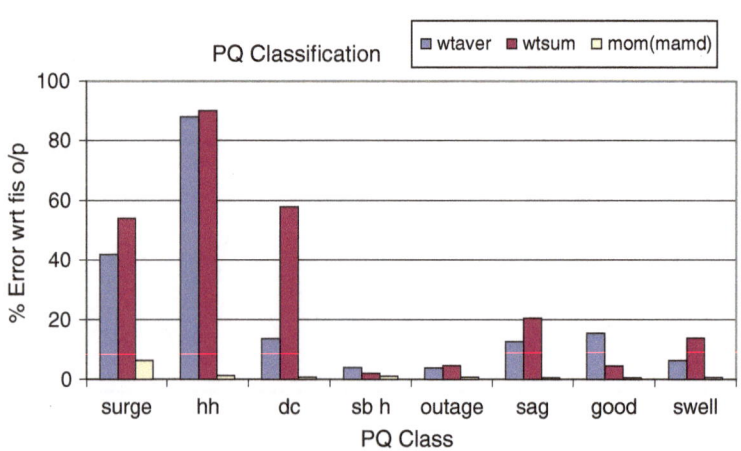

Fig. 5.19 Classification error with different defuzzification strategy (Sugeno's fuzzy inference)

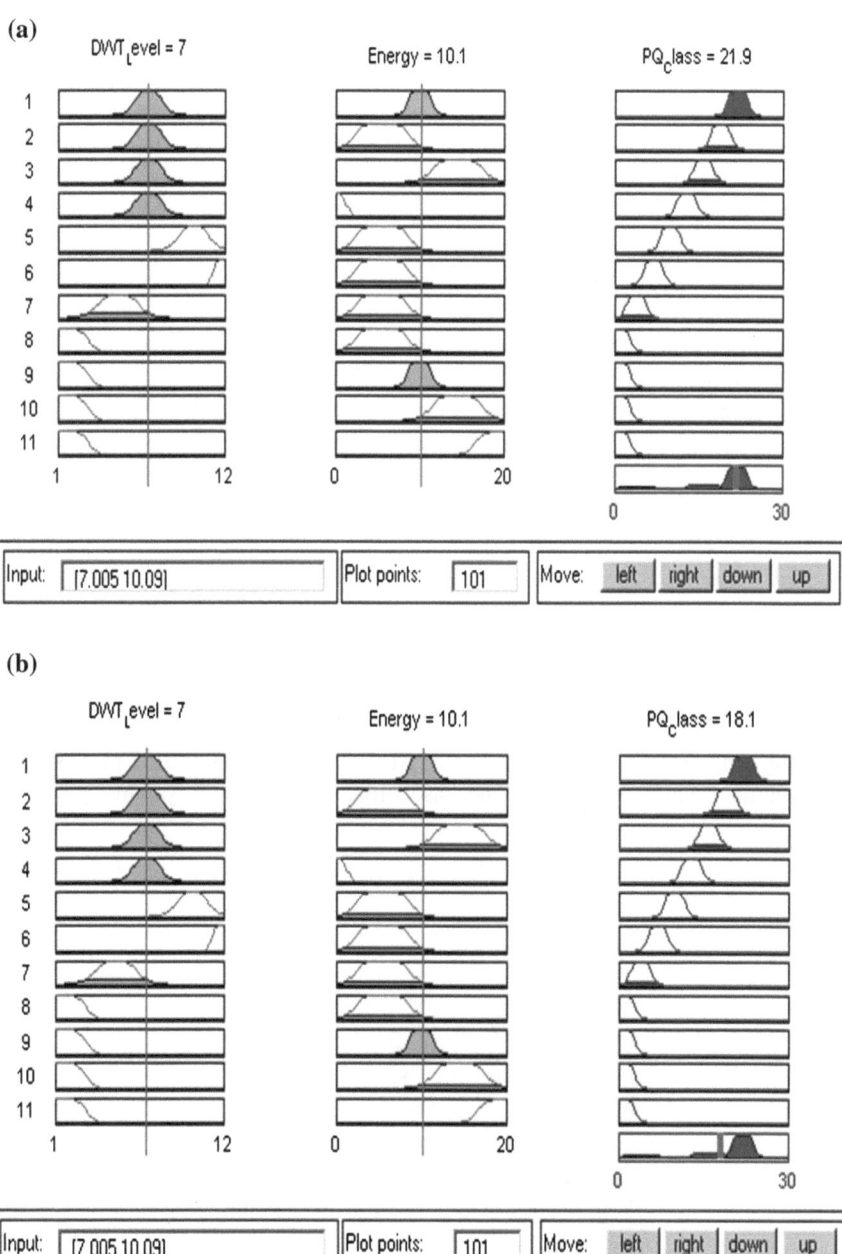

Fig. 5.20   a Rule view MoM defuzzification (Mamdani). **b** Rule view centroid defuzzification (Mamdani)

The rule base for the fuzzy decision support system is as listed in Table 5.1. The rules have been obtained in consultation with an expert power system engineer and refined after several trials. The fuzzy inferencing is done using the maximum product compositional rule of inference. If α_1, α_2, α_3, ...α_7 are the firing strength of the rule base for each category of the transient PQ disturbance (Normal, Sag, Swell, Higher Harmonic, Subharmonic, Outage, and Surge), the output is obtained as

$$\mu_0 = \alpha_1 OR\alpha_2 OR\alpha_3 OR\alpha_4 OR\alpha_5 OR\alpha_6 OR\alpha_7 = \max(\alpha_1, \alpha_2, \alpha_3, \alpha_4, \alpha_5, \alpha_6, \alpha_7) \quad (5.3)$$

Figure 5.18 plots percentage error in classification of various PQ issues by different defuzzification strategy with Mamdani's fuzzy inference method. It could be observed here that defuzzification strategy based on mean of maxima (MoM) is suitable in present application. The reason being that when the MoM strategy is used, the performance of fuzzy logic–based classifier is similar to that of a multilevel relay system [105]. As the input signal has already been preprocessed by multiresolution technique, hence input to FLC is distinguished energy levels at different levels.

Figure 5.19 plots percentage error in classification of various PQ issues by two-defuzzification strategy, that is, *weighted average* and *weighted sum* with Sugeno's fuzzy inference method. It once again proves suitability of MoM defuzzification strategy over other. With reference to *weighted average* and *weighted sum* type defuzzification strategy, the former one is more suitable in present application.

Figure 5.20 plots rule view of centroid- and MoM-based defuzzification scheme strengthening classification capability of MoM method. Figure 5.21 plots surface view of the Mamdani- and Sugeno-based fuzzy inferencing systems. From the

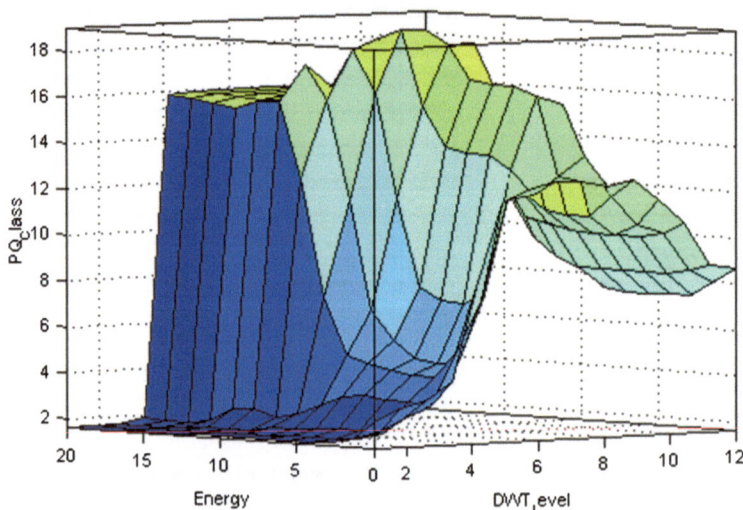

Fig. 5.21 a Surface view of the fuzzy inference systems Mamdani inferencing. **b** Surface view of the fuzzy inference systems Sugeno inferencing

Table 5.2 PQ classification; output of model and corresponding PQ problem

PQ class	Mamdani inferencing					Sugeno inferencing	
	Defuzzification methods					Defuzzification method	
	CoA	Bis	MoM	LoM	SoM	wtaver	wtsum
Surge	2.02	1.5	0.9	1.5	0.3	2.27	0.73
Higher harmonic	7.17	5.1	4.05	6.0	2.1	11.5	18.4
DC	7.43	7.2	6.9	7.5	6.3	7.95	2.95
Subharmonic	10.4	10.2	9.9	12	7.8	10.0	10.2
Outage	11.7	12.6	13.1	14.4	11.7	12.5	13.6
Swell	17.2	16.8	16.1	17.4	14.7	17	18.2
Sag	15.7	18.3	19.1	20.4	17.7	16.6	22.9
Normal	18.6	21.0	22.1	22.5	21.6	18.6	21.0

Table 5.2, it is observed that each category of waveform is successfully classified as the output from the fuzzy expert system and shows truth value of particular class that suddenly rises in most cases in comparison with normal waveform. Hence, it is obvious that the proposed approach is computationally simple in comparison with ANN-based and fuzzy expert system with Fourier linear combiner–based approaches [54, 65, 119].

5.4 Conclusions

The chapter presents a new approach for the classification of PQ disturbances using a discrete wavelet transform and fuzzy inference system. The DWT is used to extract information from the distorted signal under investigation. It separates PQ problems that overlap in time and frequency. Using the localization property gained from the finer resolution levels and the feature extraction curve, a time–frequency picture of the distorted signal is generated. The proposed feature extraction curve presents a simple classification tool for the operator to detect, localize, and classify different PQ problems. Further, the feature extraction curve is used as input to the fuzzy expert system for classification of different PQ problems. A comparative analysis of different defuzzification scheme shows the superiority of MoM method over other defuzzification methods.

Chapter 6
Conclusions and Future Directions

Abstract In this chapter, brief summaries of the chapters are presented sequentially. The organization of chapter is as follows. Concluding remarks are presented in Sect. 6.1, and possible future directions are brought out in Sect. 6.2.

Keywords DWT · PVM · GCV · CoA

6.1 Concluding Remarks

In brief, a general survey of wavelet transform has been presented in Chap. 1.

Chapter 2 presents mathematical foundation of filter bank theory. The popular pyramid structure DWT and new parallel filter DWT are introduced in detail.

Chapter 3 presents a thorough investigation of DWT from finite precision viewpoint. Theoretical formulation for prediction of errors due to finite word length hardware is laid down.

Chapter 4 presents another implementation issue related to DWT. In this, implementation of DWT has been carried out using a cluster of connected workstations in parallel virtual machine environment (PVM). The advancement in computer network technology has made availability of high-speed network quite common. Consequently, a distributed computer system has the potential to parallelize the DWT algorithm for signal processing applications. The experimental result validates effectiveness of the proposed method. The GCV-based threshold selection has been found to perform better in PVM environment than in the popular *Universal* and *Sure* method. The higher computation demand of GCV is adequately met by PVM. The suitability of parallel filter DWT structure has been proved experimentally.

Chapter 5 presents a new approach for the classification of power quality (PQ) disturbances using DWT and fuzzy inference system. In this, the DWT works as

K. K. Shukla and A. K. Tiwari, *Efficient Algorithms for Discrete Wavelet Transform*, SpringerBriefs in Computer Science, DOI: 10.1007/978-1-4471-4941-5_6, © K. K. Shukla 2013

extractor of information from signal under investigation. It separates PQ problems that overlap in time and frequency. Generation of a feature extraction curve and a time–frequency picture of the distorted signal is presented. The proposed feature extraction curve presents a simple classification tool for the operator to detect, localize, and classify different PQ problems. Further, the feature extraction curve is used as input to the fuzzy expert system for classification of different PQ problems. A comparative analysis of different defuzzification scheme shows the superiority of *mean of maxima* (MoM) method over other defuzzification methods like centroid of area (CoA), bisector of area, etc.

6.2 Future Research Directions

- To study the effect of finite word length for two-dimensional DWT (images).
- To formulate analytical expression for λ (gain parameter of smooth thresholding rules) and to prove analytically that threshold function with respect to λ is unimodel.
- Optimization of the modified thresholding function jointly with respect to gain parameter λ and threshold parameter t.
- Parallel implementation of DWT on other parallel architectures.

Bibliography

1. Abramovich F., & Benjamini, Y. (1996). Adaptive thresholding of wavelet coefficients. *Elsevier Computational Statistics and Data Analysis, 22,* 351–361.
2. Abramovich, F., Benjamini, Y., Donoho, D.L. & Johnstone, I.M. (2000). Adapting to unknown sparsity by controlling the false discovery rate, *Technical Report* 00-19, Department of Statistics, Stanford University, USA.
3. Abramovich, F., Sapatinas, T. & Silverman, B.W., Wavelet thresholding via a Bayesian approach. *Journal of the Royal Statistical Society, 60,* 725–749.
4. Andra K., Chakrabarti C., Acharya T. (2002). A VLSI architecture for lifting-based forward and inverse wavelet transform. *IEEE Transactions on Signal Processing, 50*(4), 966–977.
5. Angrisani, L., Daponte P., D'Apuzzo, M., & Testa, A. (1998). A measurement method based on the wavelet transform for power quality analysis. *IEEE Transactions on Power Delivery, 12*(4), 990–998.
6. Antoniadis, A., Bigot, J. & Sapatinas, T. (2001). Wavelet estimators in nonparametric regression: Description and simulative comparison, *Technical report IMAG*, pp. 1–78.
7. Barron, A. (1993). Universal approximation bounds for superposition of a sigmoid function. *IEEE Transactions on Information Theory, 39* (3), 930–945.
8. Bopardikar, A. S., Rao Raghuveer, M., & Adiga, S. B. (2000). Matched sampling systems, relation to wavelets and implementation using PRCC filter banks. *IEEE Transactions on Signal Processing, 48*(8), 2269–2278.
9. Bradley, J., Brislawn, C., & Hopper, T. (1993). The FBI wavelet/scalar quantization standard for gray-scale fingerprint image compression, *Proceedings of SPIE* (Vol. 1961).
10. Breveglieri, L., Dadda, L. (1998). A VLSI inner product macro cell. *IEEE Transactions on VLSI Systems, 6* (2), 292–298.
11. Breveglieri, L., Piuri, V. & Swartzlander, E.E. (1998). A serial wavelet transform processor. *Proceedings of CESA'98*, Hammamet, Tunisia (pp. 306–311), 1–4 April 1998.
12. Bruce, A. & Gao, H. Y. (1996). Understanding wave shrink: Variance and bias estimation. *Biometrika, 83,* 727–745.
13. Bruce, A., Donoho, D., & Gao, H. Y. (1996). Wavelet analysis. *IEEE Spectrum, 33,* 26–35.
14. Buckheit, J.B, Chen,S., Donoho, D.L., Johnstone, I.M & Scargle, J. (1995). *About WaveLab*: *Technical Report*. Department of Statistics, Stanford University, USA. Available on line http://www-stat.stanford.edu/wavelab.
15. Buhler, J., Shokrollahi, M. A., & Stemann, V. (2000). Fast and precise Fourier transforms. *IEEE Transactions on Information Technology, 46*(1), 213–228.

16. Burrus, C. S., McClellan, J. H., Oppenheim, A. V., Parks, T. W., Schafer, R. W., & Schuessler, H. W. (1994). Computer-based exercises for signal processing using MATLAB. Englewood Cliffs: Prentice Hall.

17. Cetin, A. E., Ansari, R. (1994). Signal recovery from wavelet transform maxima. *IEEE Transactions on Signal Process*ing, *42*(1), 194–196.

18. Chakrabarti, C., Vishwanath, M. (1995). Efficient realizations of the discrete and continuous wavelet transforms: from single chip implementations to mappings on SIMD array computers. *IEEE Transactions on Signal Process*ing, *43*(3), 759–771.

19. Chambolle, A., Devore, R. A., Lee, N.Y., & Lucier, B. J. (1998). Nonlinear wavelet image processing: variational problems, compression and noise removal through wavelet shrinkage. *IEEE Transactions on Image Processing, 7*, 319–335.

20. Chang, Y. N., Satyanarayana, J. H., Parhi, K.K. (1998). Systematic design of high-speed and low-power digit-serial multipliers. *IEEE Transactions on Circuits Systems II*: *Analog Digital Process*ing, *45*(12), 1585–1595.

21. Chang, S. G., Yu, B., & Vettreli, M. (2000). Adaptive wavelet thresholding for image denoising and compression. *IEEE Transactions on Image Processing, 9*, 1532–1546.

22. Chapa, J. O., & Rao, R. R. (2000). Algorithms for designing wavelets to match a specified signal. *IEEE Transactions Signal Processing, 48*(12), 3395–3405.

23. Chen, W. K. (Ed.). (1995). The circuits and filters handbook In: Wavelet transforms (pp. 134–219). Boca Raton: CRC Press Inc.

24. Chen, T., Vaidyanathan, P.P. (1993). Recent developments in multidimensional multirate systems. *IEEE Transactions on Circuits Systems for Video Technology*m, *3*(2), 116–137.

25. Cherkassky, V., Shao, X., Mulier, F., & Vapnik, V. (1999). Model complexity control for regression using VC generalization bounds. *IEEE Transactions on Neural Networks, 10*(5), 1075–1089.

26. Chong, U. (1996). Finite word length effects in sub-band filter banks. *ICSPAT*, 97, 604–608.

27. Chui, C. K. (1992). An introduction to wavelets. San Diego: Academic Press.

28. Cohen, L. (1989). Time-frequency distribution a review. *Proceedings of IEEE, 77*(7), 941–981.

29. Coifman, R., & Wickerhauser, M.V. (1992). Entropy-based algorithms for best-basis selection. *IEEE Transactions Information Theory, 38*, 713–718.

30. Crouse, M.S., Nowak, R.D. & Baraniuk, R.G. (1998). Wavelet-based statistical signal processing using hidden Markov models. *IEEE Transactions Signal Processing, 46*, 886–902.

31. Cvetkovic, Z. (2000). On discrete short time Fourier analysis. *IEEE Transactions on Signal Processing*, 48(9), 2628–2640.

32. Dash, P. K., Mishra, S., Salma, M. M. A., & Liew, A. C. (2000). Classification of power system disturbances using a fuzzy expert system and a Fourier linear combiner. *IEEE Transactions on Power Delivery, 15*(2), 472–477.

33. Daubechies, I. (1988). Orthonormal bases of compactly supported wavelets. *Communications on Pure and Applied Mathematics, 41*, 909–996.

34. Daubechies, I. (1990). The wavelet transform, time/frequency location and signal analysis. *IEEE Transactions on Information Theory, 36*, 961–1005.

35. Daubechies, I. (1992). Ten lectures on wavelets, *CBMS-NSF Regional Conference Series, SIAM*, Philadelphia, PA, U.S.A.

36. Daubechies, I., Grossmann, A., & Meyer, Y. (1986). Painless non-orthogonal expansions. *Journal of Mathematical Physics, 27*(5), 1271–1283.

37. Daubechies, I., Mallat, S., & Willsky, A. S. (1992). Introduction to the special issue on wavelet transforms and multiresolution signal analysis. *IEEE Transactions on Information Theory, 38*(2), 529–531.

38. Denk, T.C., Parhi, K.K. (1997). VLSI architectures for lattice structure based orthonormal discrete wavelet transforms. *IEEE Transaction Circuits Systems*: *Analog Digital Signal Process*ing, *44* (2), 129–132.

39. Donoho, D. L. (1995). Denoising by soft thresholding. *IEEE Transactions on Information Theory*, *41*(3), 613–627.
40. Donoho, D. L., & Johnstone, I. M. (1994). Ideal denoising in an orthonormal basis chosen from a library of bases, *Technical Report 461*. Department of Statistics, Stanford University.
41. Donoho, D.L., & Johnstone, I.M. (1998). Minimax estimation via wavelet shrinkage. *The Annals of Statistics*, *26*, 879–921.
42. Donoho, D. L., Vetterli, M., Devore, R. A., & Daubechies, I. (1998). Data compression and harmonic analysis. *IEEE Transactions on Information Theory*, *44*, 2435–2476.
43. Driankov, D., Hellendoorn, H. & Reinfrank, M. (1997). An introduction to fuzzy control. New Delhi: Narosa Publishing House.
44. Dugan, R. C., McGranaghan, M. F., Beaty, H.W. (1996). Electric power systems quality. New York: McGraw-Hill.
45. Evangelista, G. (1994). Comb and multiplexed wavelet transforms and their applications to signal processing. *IEEE Transactions on Signal Processing*, *42*(2), 292–303.
46. Field, D. J. (1994). What is the goal of sensory coding? *Neural Computation*, *6*(4), 559–601.
47. Forsythe, G.E., Malcolm, M. A., & Moler, C.B. (1976). Computer methods for mathematical computations. Englewood Cliffs: Prentice Hall.
48. Franklin, G. F., Powell, J. D., & Workman, M. L. (1990). Digital control of dynamic systems (2nd ed.). Reading: Addison-Wesley Publishing Company.
49. Fridman, J. & Manolakos, E. S. (1994). Distributed memory and control VLSI architectures for the 1-D discrete wavelet transform. *IEEE VLSI Signal Processing*, *VII*, 388–397.
50. Fridman, J., Manolakos, E. S. (1994). On the synthesis of regular VLSI architectures for the 1-D discrete wavelet transform, *Proceedings of the SPIE conference on mathematical imaging: wavelet applications in signal and image processing II*, San Diego CA, July 1994.
51. Fridman, J., Manolakos, E. S. (1997). Discrete wavelet transform: Data dependence analysis and synthesis of distributed memory and control array architecture. *IEEE Transactions on Signal Processing*, *45*(5), 1291–1308.
52. Fridman, J., York, B., & Manolakos, E.S. (1995). Discrete wavelet transform algorithm on the MasPar MP-1, *Proceedings of ICSPAT*, Boston.
53. Gadre, V. M. & Patney, R. K. (1994). Some novel multirate archtectures for filter realisation with reduced multiplicative complexity. *IEEE Transactions on Signal Processing*, *42*(9), 2492–2495.
54. Ghosh, A. K., & Lubkeman, D. L. (1995). The classification of power system disturbances waveforms using a neural approach. *IEEE Transactions on Power Delivery*, *10*(1), 109–115.
55. Giest, Al., et al. (1994). PVM: parallel virtual machine—a users' guide and tutorial for networked parallel computing, Cambridge: MIT Press.
56. Goyal, V. K. (2001). Theoretical foundations of transform coding. *IEEE Signal Processing Magazine*, *18*, 9–21.
57. Graps, A. (1995). Introduction to wavelets. *IEEE Computational Science and Engineering*, *2*, 50–61.
58. Gray, R. M., & Neuhoff, D. L. (1998). Quantizaion. *IEEE Transactions on Information Theory*, *44*, 2325–2383.
59. Grossmann, A., & Morlet, J. (1984). Decomposition of hardy functions into square integrable wavelets of constant shape. *SIAM Journal of Mathematics Analysis*, *15*, 723–736.
60. Grzeszczak, A., Mandal, M.K., Panchanathan, S. (1996). VLSI implementation of discrete wavelet transform. *IEEE Transactions on VLSI Systems*, *4*, 421–433.
61. Hanselmann, H. (1987). Implementation of digital controllers—a survey. *Automatica*, *23*(1), 7–32.
62. Harish, K. S., & Prabhu, K. M. M. (2000). Fixed point error analysis of two DCT algorithms. *Proceedings of IEE Image Signal Processing*, *147*(2), 131–137.
63. Harris, F. J. (1978). On the use of windows for harmonic analysis with the discrete Fourier transform. *Proceedings of IEEE*, *66*(1), 51–83.

64. Herley, C., Vetterli, M. (1993). Wavelets and recursive filter banks. *IEEE Transactions on Signal Processing, 41*(8), 2536–2556.
65. Heydt, G. T., & Gali, A. W. (1997). Transient power quality problems analyzed using wavelets. *IEEE Transactions on Power Delivery, 12*(2), 908–915.
66. Hosur, S., & Tewfik, A. H. (1997). Wavelet transform domain adaptive FIR filtering. *IEEE Transactions on Signal Processing, 45*(3), 617–630.
67. Jackson, L.B. (1993). Digital filters and signal processing (2nd ed.). Norwell, Massachusetts: Kluwer Academic Publishers, Seventh Printing.
68. Jansen, M., et. al. (1997). Generalized cross validation for wavelet thresholding. *Elsevier Signal Processing, 56*(1), 33–44.
69. Jayant, N., Johnston, J., & Robert, S. (1993). Signal compression based on models of human perception, *Proceedings of IEEE, 81*(10), 1385–1422.
70. Johnstone, I.M., & Silverman, B.W. (1997). Wavelet threshold estimators for data with correlated noise, *Journal of the Royal Statistical Society, 59*, 319–351.
71. Kaiser, G. (1994). A friendly guide to wavelets. *Birkhauser*, Boston, MA, U.S.A.
72. Kandil, N., et al. (1992). Fault identification in an AC-DC transmission system using neural networks. *IEEE Transactions on Power System, 7*(2), 812–819.
73. Kim, H. J., Li, C. C. (1998). Loss less and lossy image compression using biorthogonal wavelet transforms with multiplier less operations. *IEEE Transactions on Circuits Systems II: Analog Digital Processing, 45*(8), 1113–1118.
74. Kim J.T., Lee Y.H., Isshiki T., Kunieda H., "Scalable VLSI architectures for lattice structure-based discrete wavelet transform. *IEEE Transaction Circuits Systems II: Analog Digital Processing, 45*(8), 1031–1043.
75. Koeck, P. J. B. (2001). Quantization errors in averaged digitized data. *Signal Processing, 81*, 345–356.
76. Koornwinder, T. H. (Ed.) (1993). Wavelets: an elementary treatment of theory and applications. River Edge: World Scientific.
77. Kulkarni, S., Gadre, V. M., & Bellary, S. V. (2000). Nonuniform M-band wavepacketes for transient signal detection. *IEEE Transactions on Signal Processing, 48*(6), 1803–1806.
78. Lang, R., Plesener, E., Schroder, H., & Spray, A. (1994). An efficient systolic architecture for the one dimensional wavelet transform: *Proceedings of SPIE conference on wavelet applications* (pp. 925–935). Orlando, April 1994.
79. Lang, M., Guo, H., Odegard, J. E., Burrus, C. S., & Wells, Jr R. O. (1996). Noise reduction using an un-decimated discrete wavelet transform. *IEEE Signal Processing Letters, 3*, 10–12.
80. Lee, C. C. (1990). Fuzzy Logic in control systems: Fuzzy logic controller—part I and II, *IEEE Transactions on Systems, Man and Cybernetics*, 20(2), 404–435.
81. Lee, D. T. L., & Yamamoto, A. (1994). Wavelet analysis: theory and applications. *Hewlett-Packard Journal, 45*(6), 44–54.
82. Lim, Y. C., Sun, Y., & Jun, Y. (2002). Design of discrete-coefficient FIR filters on loosely connected parallel machines. *IEEE Transactions on Signal Processing, 50*(6), 1409–1416.
83. Liu, B. (1971). Effect of finite word length on the accuracy of digital filters—a review, *IEEE Transaction Circuit Theory, CT-18*, 670–677.
84. Ma, Y. (1997). An accurate error analysis model for fast Fourier transform. *IEEE Transactions on Signal Processing, 45*(6), 1641–1645.
85. Ma, J., Pahri, K. K., & Deprettere, F. (2001). A unified algebraic transformation approach for parallel recursive and adaptive filtering and SVD algorithm. *IEEE Transactions on Signal Processing, 49*(2), 424–437.
86. Mallat, S.G. (1989). Multi frequency channels decompositions of images and wavelet models. *IEEE Transactions on Acoustics, Speech, and Signal Processing, 37*(12), 2091–2110.

87. Mallat, S. G. (1989). A theory of multiresolution signal decomposition: The wavelet representation. *IEEE Transactions on Pattern Analysis and Machine Intelligence, 1*(7), 674–693.
88. Mallat, S. G., Zhong, S. (1992). Characterization of signals from multiscale edges. IEEE Transactions on pattern analysis and Machine Intelligence, *14*(7), 710–732.
89. Mamdani, E.H., & Assilian, S. (1975). An experiment in linguistic synthesis with a fuzzy logic controller. *International Journal of Man-Machine Studies, 7*(1), 1–13.
90. Mandal, M.K., Panchanathan, S. & Aboulnasr, T. (1996). Choice of wavelets for image compression. *Lecture Notes in Computer Science* (Vol. 1133, pp. 239–249). Springer: Berlin.
91. Marcos, M.M., & Ibrahim, A. (1999). Electric power quality and artificial intelligence: Overview and applicability. *IEEE Spectrum, 19*(6), ISSN (0272-1724).
92. Marron, J. S., Adak, S., Johnstone, I. M., Neuman, M. H., & Patil, P. (1998). Exact risk analysis of wavelet regression. *Journal of Computational and Graphical Statistics, 7*, 278–309.
93. Mathworks Inc. Simulink 3.0. http://www.mathworks.com/products/simulink/.
94. Mathworks Inc. Matlab 5.3. http://www.mathworks.com/products/matlab/.
95. McEachern, A. (1988). Handbook of power signatures. *Basic Measuring Instruments*, Foster City, CA.
96. Mertins, A., & Karp, T. (2002). Modulated, perfect reconstruction filter banks with integer coefficients. *IEEE Transactions on Signal Processing, 50*(6), 1398–1408.
97. Meyer, Y., & Raphael, L. (Eds.). (1992). Wavelets and their applications. Boston: Jones and Bartlett Publishers.
98. Middleton, R., & Goodwin, G. (1990). Digital control and estimation—a unified approach. Prentice Hall: Englewood Clifs.
99. Mitra, S. K. (1998). Digital signal processing: A computer based approach. New Delhi: Tata McGraw Hill Edition.
100. Moler, C. (1996). Floating points: IEEE Standard unifies arithmetic model, *Cleve's Corner, the MathWorks, Inc.*, http://www.mathworks.com/company/newsletter/clevescorner/.
101. Motard, R. L., & Joseph, B. (Eds.). (1994). Wavelet applications in chemical engineering. Boston: Kluwer Academic Publishers.
102. Nason, G.P. "Wavelet regression by cross-validation", *Technical Report* 447, 1994, Department of Statistics, Stanford University, USA.
103. Newland, D. E. (1993). An introduction to random vibrations, spectral and wavelet analysis. *Longman scientific and technical* (3rd Ed., pp. 295-370), London, New York: Wiley.
104. Ogata, K. (1995). Discrete-time control systems (2nd ed.). Prentice Hall: Englewood Cliffs.
105. Oppenheim, A. V., & Schafer, R. W. (1993). Digital signal processing. Englewood Cliffs: Prentice-Hall.
106. Oppenheim, A. V., Weinstein, C. J. (1972). Effects of finite register length in digital filtering and fast Fourier transform. *Proceedings of the IEEE* (*Invited Paper*), *60*, 957–976.
107. Oppenheim, A.V., Schafer, R. W., & Buck, J. R. (1999). Discrete time signal processing. New Jersey: Prentice Hall.
108. Parhi, K. K., Nishitani, T. (1993). VLSI architectures for discrete wavelet transforms. *IEEE Transactions on VLSI Systems, 1*(2), 191–202.
109. Pillay, P., & Bhattacharjee, A. (1996). Application of wavelets to model short-term power system disturbances. *IEEE Transactions on Power Systems, 11*(4), 2031–2037.
110. Rao, R. M., Bopardikar, A. S. (1998). Wavelet transforms: Introduction to theory and applications. Boston: Addison Wesley.
111. Resnikoff, H. L., & Burrus, C. S. (1990). Relations between the Fourier transform and the wavelet transform. Proceeding of SPIE—The International Society for Optical Engineering, *1348*, 291–300.
112. Resnikoff, H. L., & Wells, R. O. (1998). Wavelet analysis: The scalable structure of information. ISBN: 3-540-780-76-9, New York: Springer.

113. Rioul, O., & Duhamel, P. (1992). Fast algorithms for discrete and continuous wavelet transforms. *IEEE Transactions on Information Theory, 38*, 569–586.
114. Rioul, O, & Vetterli, M. (1991). Wavelets and signal processing, *IEEE Signal Processing Magazine*, pp. 14–38.
115. Roberts, R.A., & Mullis, C.T. (1987). Digital signal processing. Reading: Addison-Wesley Publishing Company.
116. Roy, P. K. (2002). Emerging trends in parallel computing: The high performance clusters. *CSI Communications*, pp. 4–7.
117. Santoso, S., Powers E. J., Grady W., & Parsons, A. (2000) Power quality disturbance waveform recognition using wavelet-based neural classifier, part 2: Application. *The 1997 IEEE/PES Winter Meeting*, New York, NY, U.S.A.
118. Santoso, S., Powers, E. J., & Hofman, P. (1996). Power quality assessment via wavelet transform analysis. *IEEE Transactions on Power Delivery, 11*(2), 924–930.
119. Santoso, S., Powers, E. J., Grady, W., & Parsons, A. (2000) Power quality disturbance waveform recognition using wavelet-based neural classifier, part 1: Theoretical foundation. *The 1997 IEEE/PES Winter Meeting*, New York, NY, U.S.A.
120. Sardy, S, Tseng, P., & Bruce, A. (2001). Robust wavelet denoising. *IEEE Transactions on Signal Processing, 49*(6), 1146–1152.
121. Shensa, M.J. (1992). The discrete wavelet transform: Wedding the a trous and Mallat algorithms. *IEEE Transactions on Signal Processing, 40*(10), 2464–2481.
122. Souani C., Atri, M., Abid, M., Torki, K. & Tourki, R. (2000). Design of new optimized architecture processor for DWT. *Real Time Imaging, 6*(4), 297–312.
123. Souani, C. et al. (2000). VLSI design of 1-D DWT architecture with parallel filters. *Integration, the VLSI Journal, 29*, 181–207.
124. Stanhill, D., Zeevi, Y. Y. (1998). Frame analysis of wavelet type filter banks. *Signal Processing, 67*, 125–139.
125. Stein, C. (1981). Estimation of the mean of a multivariate normal distribution. The Annals of Statistics, *9*, 1135–1151.
126. Strang, G., & Nguyen, T. (1996). Wavelets and filter banks. Wellesley: Wellesley-Cambridge Press.
127. Strange, G. (1989). Wavelet and dilation equations: a brief introduction. *SIAM Review, 31*(4), 614–627.
128. Strange, G. (1993) Wavelet transforms versus Fourier transforms, *Bulletin of the American Mathematical Society, 28*(2), 288–305.
129. Sugeno, M. (1985). Industrial applications of fuzzy control. Amsterdam: Elsevier.
130. Sweldens, W. (1996). Wavelets: What next? *Proceedings of the IEEE, 84*(4), 680–685.
131. Unser, M. (2000). Sampling—50 years after Shannon. *Proceedings of the IEEE, 88*(4), 569–587.
132. Unser, M., Aldroubi, A., & Schiff, S. J. (1994). Fast implementation of the continuous wavelet transform with integer scales. *IEEE Transaction SP, 42*(2), 3519–3523.
133. Usevitch, B. E. & Betancourt, C. L. (1999). Fixed point error analysis of two channel perfect reconstruction filter banks with perfect alias cancellation. *IEEE Transactions on Circuits and Systems II, 46*(11), 1437–1440.
134. Vaidyanathan, P. P. (1993). Multirate systems and filter banks. Prentice Hall PTR: Englewood Cliffs.
135. Vaidyanathan, P. P. (2001). Generalizations of the sampling theorem: Seven decades after Nyquist. *IEEE Transactions on Circuits and Systems—I, 48*(9), 1094–1109.
136. Vaidyanathan, P. P., & Djokovic, I. (1995). Wavelet transforms. In W.K. Chen (Ed.), *The circuits and filters handbook* (pp. 134–219). Boca Raton: CRC Press.
137. Verdu, S., Fifty years of Shannon theory. *IEEE Transactions on Information Theory, 44*(6), 2057–2078.
138. Vettereli, M. (2001). Wavelets, approximation and compression. *IEEE Signal Processing Magazine, 18*, 59–73.

139. Vetterli, M., Herley, C. (1992). Wavelets and filter banks: Theory and design. IEEE Transactions on Signal Processing, *40*(9), 2207–2231.
140. Vetterli, M., & Kovacevic, J. (1995). Wavelets and subband coding. Englewood Cliffs: Prentice Hall.
141. Vishwanath, M. (1994). The recursive pyramid algorithm for the discrete wavelet transform. IEEE Transactions on Signal Processing, *42*(3), 673–676.
142. Vishwanath M., Owens R.M., Irwin M.J. (1995). VLSI architectures for the discrete wavelet transform. *IEEE Transactions on Circuits Systems II: Analog Digital Processing, 42*(5), 305–316.
143. Wang, Z., & Bovik, A. C. (2002). A universal image quality index. *IEEE Signal Processing Letter, 9*(3), 81–84.
144. Weinstein, C., & Oppenheim, A.V. (1969). A comparison of round off noise in floating point and fixed point digital filter realizations, *Proceedings of the IEEE (Letters), 57,* 1181–1183.
145. Williams, J. R., & Amaratunga, K. (1994). An introduction to wavelets in engineering. *International Journal for Numerical Methods in Engineering, 37,* 2365–2388.
146. Yarlagadda, R., & Hershey, J E. (1987). Signal processing general. *Encyclopedia of Physical Science and Technology* (Vol. 12, pp. 626–646), Academic Press.
147. You, J. (1999). Distributed system design. Boca Raton: CRC.
148. You, J., & Bhattacharya, P. (2000). A wavelet based coarse-to-fine image matching scheme in a parallel virtual machine environment. *IEEE Transactions on Image Processing, 9*(9), 1547–1559.
149. Zeng, Y., Cheng, L., Bi, G., & Kot, A. C. (2001). Integer DCTs and fast algorithms. *IEEE Transactions on Signal Processing, 49*(11), 2774–2782.
150. Zervakis, M. E., Sunderarajan, V., & Parhi, K. K. (2001). Vector processing of wavelet coefficients for robust image denoising. *Elsevier Image and Vision Computing, 19,* 435–450.
151. Zhang, X. P. (2001). Thresholding neural network for adaptive noise reduction. *IEEE Transactions on Neural Network, 12*(3), 567–584.
152. Zhu, Y., Zhou H., Gu, H., & Wang, Z. (1999). Fixed point error analysis and an efficient array processor design of two dimensional sliding DFT, *Signal Processing, 72,* 191–201.